Northern Hemisphere Edition

THE ULTIMATE BEGINNER TELESCOPE GUIDEBOOK

Over 250 Targets for Beginners

Plus More Than 25 Smart Telescopes Targets

John A. Read

Host of *Learn To Stargaze* on YouTube

www.LearnToStargaze.com
facebook.com/LearnToStargaze
Instagram: @LearnToStargaze

Star maps used in this book were created by Kovalik Media. The star map source is Stellarium (https://stellarium.org), an open-source stargazing program.

Executive Editor/Consultant: David M. F. Chapman
Content Editors: Dr. Scott Mellis, Paul Clarke
Interior Design by Stellar Publishing

For bulk or custom orders, visit: www.learntostargaze.com/contact

8.5x8.5 Paperback ISBN: 978-1738659029
8.5x8.5 Hardcover ISBN: 978-1738659043

To find the astronomy club nearest you, please visit:

https://nightsky.jpl.nasa.gov (USA)

https://www.rasc.ca/locations-across-canada (Canada)

https://www.skyandtelescope.com/astronomy-clubs-organizations (All)

This edition of the book contains several items visible only to those living north of the Equator. For those in the Southern Hemisphere, we recommend *50 Things to See with a Telescope - Southern Hemisphere Edition.* Follow Learn To Stargaze on Facebook, YouTube or Skool for updates on a Southern Hemisphere edition of this book.

This book is dedicated to my five-year-old son, Edwin.

May you grow to love the sky.

Dedication from *50 Things to See with a Small Telescope* (2012)

I would like to express my gratitude to Marni Berendsen, developer of the NASA Night Sky Network, for her fantastic contribution of editing and fact-checking the original version of this book: *50 Things to See with a Small Telescope* (2012). This self-published book kicked off my professional "space" journey, enabling me to work on these books full time, leading to a degree in Astrophysics from SMU and a Masters in Space Systems Engineering from Johns Hopkins.

I would also like to thank the Mount Diablo Astronomical Society (MDAS) for feeding my desire to learn more about the Universe. This book would not have been possible without the wonderful folks at MDAS.

Table of Contents

Preface

It's hard to believe that I wrote *50 Things to See With a Small Telescope* almost fifteen years ago. I would never have guessed that small book would allow me to quit my job, study astrophysics full time, and become a professional author.

Back in 2012, I spent most of my free time volunteering with the Mount Diablo Astronomical Society in Northern California, sharing views of the night sky through telescopes at various events. I originally wrote the book for myself as a quick reference in advance of each stargazing event.

I printed and self-published the book myself on some lesser-known platforms. Eventually, I figured out how to self-publish on Amazon and was shocked when I sold about 100 copies that first Christmas in 2013. Since people were actually reading the book, I rewrote it, incorporating everything I'd learned that year. Shortly afterward, my company moved me to South Carolina, where I joined the Carolina Skygazers.

That next Christmas, 2014, everything changed, and I knew our lives would never be the same. I checked my sales during the first week in December and realized that I had sold hundreds of copies in a single day! Over the next year, I got to work improving the book even more, and even hiring translators to translate it into several languages.

After Christmas 2015, the book had earned enough that I felt confident I could make a living as a stargazing author. But I struggled with impostor syndrome. I felt I had no credibility and told myself I needed to break into the space industry somehow … some way.

We put our house on the market, and I gave myself two options. Option one: if it sold, we'd move to Washington DC, and I'd try to get a job at NASA headquarters. Option two: if the house didn't sell, we'd move our family to Canada, and I'd go back to university and complete a degree in astrophysics.

Showing off an exoplanet transit observation completed as part of my degree in astrophysics at Saint Mary's University

Well, the house didn't sell, and I was accepted back at Saint Mary's University in Halifax, Nova Scotia. So we packed up the car, left our big five-bedroom house in South Carolina, and moved our growing family (we now had two toddlers) into a small one-bedroom place in Canada.

Over the next four years, while completing my degree in astrophysics, I published seven additional titles. Not all of them were bestsellers, but I certainly learned a lot! I joined the local board of directors of the Royal Astronomical Society of Canada (RASC) and continued doing astronomy outreach as much as I could. I even got a job as a telescope operator at the Burke-Gaffney Observatory.

My book *50 Things to See with a Telescope [For] Kids* was a smashing success, holding the top astronomy book spot on Amazon for several holiday seasons in a row. More books followed. *50 Things to See on the Moon* won the Simon Newcomb Award for excellence in science communication from the RASC, and *50 Animals That Have Been to Space* won a Gold Medal from the Junior Library Guild. I partnered with Chris Vaughan to complete a massive project, *110 Things to See with a Telescope*, which gives stargazers the chance to earn their Messier Object observing certificate (Tim Russ from *Star Trek* wrote the foreword for that one).

In 2020, during the pandemic, I started the *Learn To Stargaze* YouTube channel. Two years later, I began my Master of Science in Space Systems Engineering at Johns Hopkins University. In 2023, I purchased a stargazing park near Peggy's Cove, Nova Scotia, and took over the Abbey Ridge Observatory from the late Dr. David Lane.

Now, in 2026, I have a new challenge that I hope to overcome by dramatically raising the quality of everything I do. As a consequence of popularity, several of my books have been pirated and plagiarized. My Amazon "Best Seller" spots have been taken by AI-generated knockoffs. But as author Seth Godin says, "The enemy of an author is not piracy, but obscurity."

This book aims to be better than ever. These books are as authentic as they come, and my hope is that radical authenticity will help these titles earn a place as reliable stargazing classics for years to come.

Spreading Joy with Telescopes

In the age of smart telescopes and instant gratification, I sometimes wonder whether there's still joy to be found in stargazing with a telescope, one without any electronics. Back in 2024, my friend Dr. Sian Proctor asked me to set up some telescopes at Arizona State University's *Space2Inspire* Weekend. So I set up a telescope on the rooftop bar at the Hyatt, where the event's speakers were staying. The telescope was a simple Sky-Watcher Evolux 82, on an AZ mount.

It was amazing to capture the reactions of space professionals using a small telescope for the first time.

At this event, I was reminded just how much joy a simple telescope can bring. I was inspired to keep doing what I do, writing books like this to spread the wondrous views of space through the lens of a beginner's telescope.

Sometimes we need to be reminded of the simple joy of being outside at night, under a dark sky. If we're lucky, we'll be joined by close friends, and a hot cup of coffee.

-J.R. Halifax, Nova Scotia, 2025

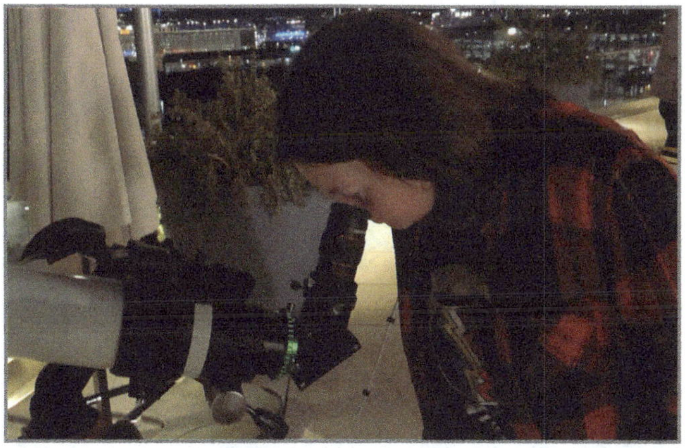

NASA astronaut Cady Coleman observes Jupiter for the first time through the Sky-Watcher Evolux.

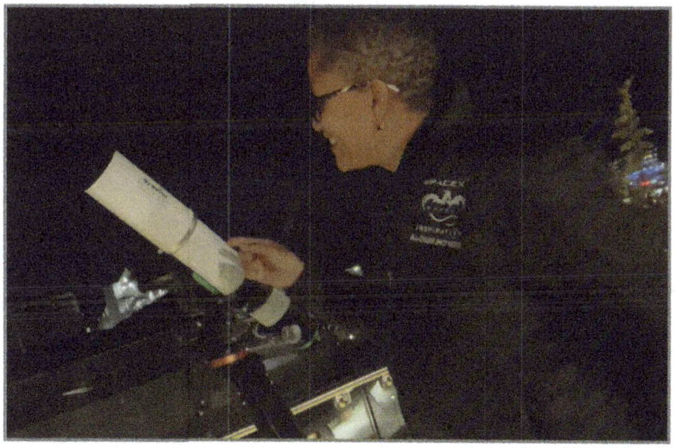

Inspiration4 mission pilot and astronaut Dr. Sian Proctor with the Sky-Watcher Evolux 82.

Introduction to the Night Sky

Targets for the Unaided Eye

Humans have been stargazing since the dawn of civilization without any technology whatsoever. A lot of people who are interested in space rush to get a telescope. However, the simple act of stargazing on any cloudless night does not require any equipment at all.

There is a lot you can see with just your eyes. If you're new to stargazing, there's a lot you're about to experience. A telescope is just a small part of the broader stargazing hobby. I've found that most stargazing events, even those that involve a telescope, consist mainly of just looking up.

I've even had some people ask me what telescope they should get to view "the constellations." As you may know, a telescope would actually provide too much magnification to view the entirety of any constellation!

The other aspect of stargazing that does not need to involve a telescope is the adventure itself, whether it's traveling to Iceland to see the Northern Lights, being under the shadow of a total solar eclipse, or hiking out to a dark sky location far from the hustle and bustle of the city. All of these activities push us outside our comfort zones and let us see the night sky in all its beauty.

There's much to see in the sky, and if you're going to spend your time looking through a telescope, I urge you first to fall in love with the night sky by spending time looking up with just your eyes. It's the way humans have done it for thousands of years, and we should continue to do it for thousands more to come.

Constellations

I live in a small city, and every night when I'm walking the dog, I look up and attempt to identify as many constellations as I can. I absolutely love the sky, and the constellations are like old friends. When I first see a summer constellation after a long Nova Scotia winter, it's like welcoming a family member I haven't seen in a while.

The brightest stars in the constellations can be seen from a wide open field in my city, but to see all the stars that form the popular star lines, it helps to get out of town.

However, the star lines are not the constellations themselves; these are just added to maps to help us locate them. Constellations are actually regions of the sky, mapping out a given territory like states on a map. The entire sky is divided into 88 of these regions.

Learning to recognize the constellations is important because they lead you to the telescope targets within. Star maps containing all the northern hemisphere constellations are found on pages 130-133.

The constellation Orion

The Milky Way

All of the stars that you see in the night sky are part of our home galaxy, the Milky Way. When we say we "see" the Milky Way, we are actually looking out through the plane of our galaxy.

If you live near a large town or city, you probably cannot see the white wisps and dark clouds of gas that make up the plane of the Milky Way. From within a large city, you can generally see only about a dozen stars.

The Milky Way contains between 300 billion and 400 billion stars! That's why it appears so brilliant from truly dark skies, far from city lights.

Explore the plane of the Milky Way by starting at one horizon and working your way across to the other. With patience, you'll find many star clusters beyond those mentioned in this book.

Milky Way from Stargaze NS

Find Dark Skies

We talk a lot about dark skies in the book, but how do you find them? The easiest way is to consult a light pollution map. However, even if you go to a dark sky location, you also need to be "Moon aware." If the Moon is in the sky and it's past the crescent phase, it will be very difficult to see the Milky Way or any other galaxies through a telescope.

Light pollution maps tend show the brightest locations, generally cities and towns, in white and red, while the suburbs are yellow and dark skies are "blue zones."

You may hear astronomers talk about the "Bortle" Scale. This is a scale from 1 (darkest) to 9 (brightest) to categorize light pollution. In general, only deserts and some national parks are Bortle 1. Locations in the countryside, far from the city, can be Bortle 2, but tend to be Bortle 3 if you can see the glow of a town's lights on the horizon. The suburbs are Bortle 4–6, while cities and towns are Bortle 7–9.

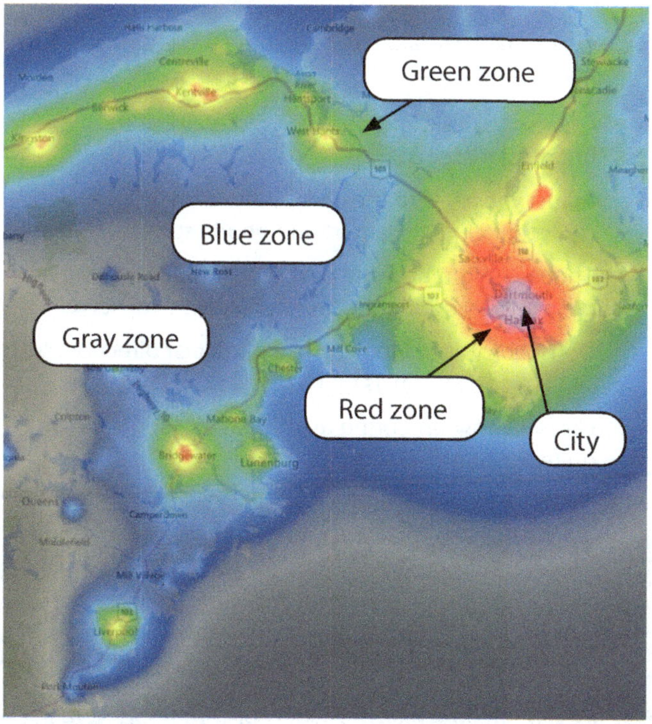

Light Pollution Map

Meteor Showers

Meteor showers occur on a predictable schedule several times per year. Comets and some asteroids leave behind trails of debris along their orbits, and when Earth passes through these swarms, some of this debris enters the atmosphere and burns up, which we see as meteors.

Some meteor showers are more impressive than others. For example, the Perseids (mid-August) and the Geminids (December) have reputations as the most impressive. Meteor showers get their names from the constellation in which the meteors appear to radiate.

Most meteors are smaller than a golf ball, but you can see them because they travel at tens of kilometres per second. When they hit the atmosphere, they generate enough heat to burn up in a bright flare.

A Geminid meteor over Big Lake, Nova Scotia.

Some tips for viewing meteor showers:

- Stay up late. Meteor showers tend to get better after midnight.
- Find dark skies. Light pollution hides all but the brightest meteors.
- Adjust your eyes to the dark by avoiding artificial light sources, including camera and phone screens.
- Choose a night near the peak where the Moon will not be in the sky during the times you plan to observe.
- Don't worry about looking in a particular direction. Just sit back and take in as much of the sky as possible.
- Dress warmly, and bring friends. Meteor showers can be quite exciting!

Meteor, Meteorite, and Meteoroid!

Even I get these terms confused. A meteoroid is the term for a small rock from space. Once the rock enters the atmosphere, the light you see is a meteor, or "shooting star." To remember this, note that we say, "meteor showers," and not "meteoroid showers." A space rock is only called a meteorite once it reaches the ground.

Meteor Shower Dates (approx.)		
Quadrantids	Early January	☐
Lyrids	Late April	☐
Eta Aquariids	Early May	☐
Delta Aquariids	Late July	☐
Perseids	Mid-August	☐
Draconids	Early October	☐
Orionids	Late October	☐
Leonids	Mid-November	☐
Geminids	Mid-December	☐

The Northern Lights

The Aurora, otherwise known as the Northern Lights, occurs when strong bursts of solar wind interact with Earth's magnetic field. Charged particles flow along Earth's magnetic field lines, emitting light in various colors from orange to green to yellow.

At high northern latitudes, such as Alaska or Norway, the Northern Lights appear frequently, but are more common during solar maximum, a period of higher solar storm activity that occurs every 11 years or so. The last solar maximum peaked in 2024, and the next peak is expected around 2035.

During geomagnetic storms, when Earth is struck by strong coronal mass ejections (CMEs) from the Sun, the Northern Lights can be seen from farther south in latitude. On May 10, 2024, I was in Tucson (32° N) at Biosphere 2, and even with a Crescent Moon above the horizon, the entire northern half of the sky turned bright pink. It stayed this way for most of the night.

Five months later, another series of CMEs passed by Earth. This time, I was at Stargaze Nova Scotia (44° N). The Aurora was so incredibly bright that we could walk by its light.

Some tips for viewing the Aurora:

- Get connected. There are groups on social media platforms, such as Facebook, where members alert each other when the Northern Lights are visible in their area.
- Find dark skies. Just like when observing the Milky Way, the Northern Lights look far better without light pollution.
- Adjust your eyes to the dark. Avoid artificial light sources, including camera and phone screens.
- Get advanced notice. Check websites such as SpaceWeather.com or Aurora-specific apps.
- Book trips to northern latitudes during solar maximum.

The Northern Lights over the Abbey Ridge Observatory

Lunar Eclipses

Lunar eclipses occur only during a Full Moon when the Moon passes into Earth's shadow. A total lunar eclipse occurs when the Moon passes fully into the umbra, where all light from the Sun is blocked. A partial lunar eclipse occurs when only part of the Moon enters the umbra. A lesser-known lunar eclipse occurs when the Moon enters only the penumbra (the indirect shadow). This is called a penumbral eclipse, but is difficult to distinguish from a regular Full Moon.

Lunar eclipses occur, on average, a couple of times per year, although in some years there may be none. It also depends on where you live, as a lunar eclipse may occur before moonrise or after moonset in your location. To determine the time of the next lunar eclipse, it's best to look it up on a website such as timeanddate.com.

Why don't lunar eclipses occur every month? The Moon's orbit is inclined relative to Earth's orbit around the Sun. The points where the Moon's orbit intersects Earth's orbital plane are called nodes, and lunar eclipses occur when the Full Moon is near one of these nodes.

Lunar eclipse

How lunar eclipses work

Solar Eclipses

Solar eclipses are among the most wondrous experiences that a human can witness in nature. Solar eclipses occur when the Moon passes in front of the Sun. Partial solar eclipses occur fairly often and are not a newsworthy event. Most of the time, they go unnoticed. If the Moon is far from Earth in its orbit, it doesn't completely cover the Sun at all; this is called an annular eclipse. However, if the Moon orbits close to Earth and passes in front of the Sun, somewhere on Earth will experience a total solar eclipse.

A total solar eclipse is worth traveling for. They occur over a given location only about once every three centuries. This is because the Moon's shadow traces a path roughly 200 kilometres wide. It's important to note that, even in the case of a 99%-covered Sun, **you are not experiencing a total eclipse**. Only when the eclipse reaches "totality—where the Sun is completely covered—do observers under its shadow experience the full majesty of the event.

As totality nears, the air cools, birds return to their nests, and insects grow louder. Even people begin to act a bit weird. They may start hooting and hollering with excitement. When the Moon fully covers the Sun, an ominous black void appears in the sky. This void is surrounded by the Sun's corona—eerie rays of light that look like fire captured in a photograph, frozen in time.

Only during totality, when the Sun is 100% blocked by the Moon, is it safe to look directly at the eclipse without eye protection. This is the time to bask in the experience with your fellow humans, feeling dwarfed by the inevitable movements of the heavenly bodies along their trajectories through the cosmos.

Not only does the sky turn nearly black as night, but bright stars and planets also become visible. The horizon glows like a sunset in all directions.

Totality lasts only a few minutes, and when it's over, the Sun returns, leaving behind memories that will last a lifetime. Everyone takes a moment to catch their breath and come down from the emotional high. After that, nearly everyone starts planning their next eclipse adventure.

Storytown Cottages, Doaktown, New Brunswick, Canada

Stargazing Basics

This diagram shows the sky from the perspective of an observer on Earth's surface. The entire sky, as observed from Earth, is known as the **Celestial Sphere**.

These are the parts of the Celestial Sphere that an astronomer might use to describe the location of an object in the sky.

For the purpose of this diagram, pretend you are living about halfway between the North Pole and the Equator and are lying on your back facing south (and you have super fish-eye vision).

While some of these explanations may seem overly simplistic, and others overly complex, they represent how these terms are generally used by seasoned stargazers.

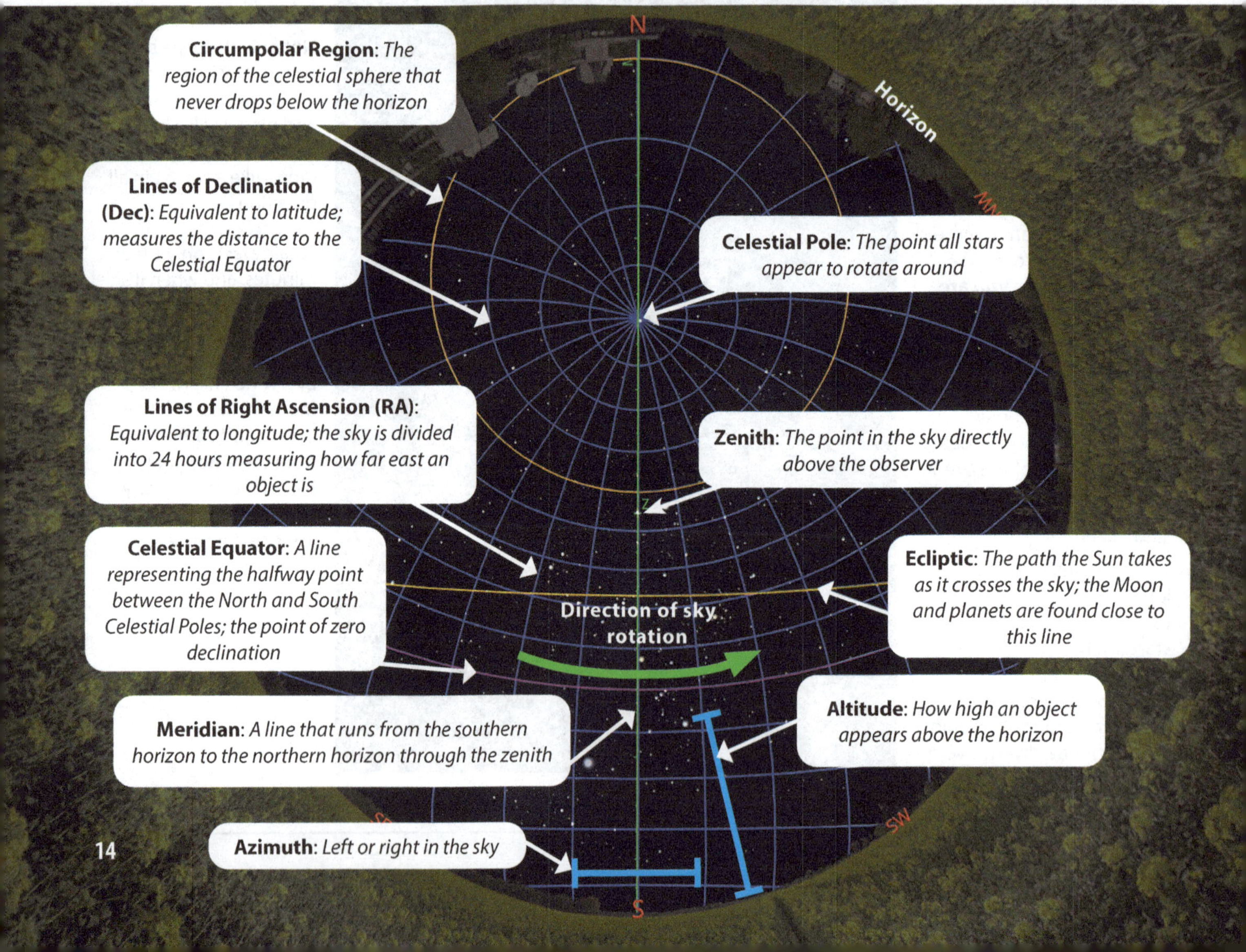

Catalogues

Most objects in space get their names from catalogues developed by astronomers over the past 250 years. These were lists of individual stars, double stars, and deep-sky objects, discovered as these astronomers carefully surveyed the night sky.

Deep-sky objects from these catalogues are where seasoned amateur astronomers spend most of their observing time. Many people start with the "Messier Catalogue," a list of 110 objects recorded by Charles Messier in the 1770s. Messier was a comet hunter, and his list was composed of objects that were not comets. Little did he know that his list would be immortalized as a compilation of some of the most interesting objects in the sky.

Seasoned amateur astronomers often pursue targets from other lists, such as the Caldwell Catalogue: a 109-item extension to the Messier Catalogue (List), that summarizes the next-best targets.

The most comprehensive list of deep-sky objects is called the New General Catalogue (NGC). This list contains 7,840 items and overlaps with both the Messier and the Caldwell Catalogues. With the advent of the camera, this list was expanded further to include items newly discovered through cameras. These items have the designation "IC" (Index Catalogue).

Messier List (M):

- Compiled by Charles Messier (1730–1817), a French comet hunter.
- Mainly deep-sky objects (not comets). Includes a double star and an asterism.
- 110 "M" objects (103 original, 7 added later). Listed in order of discovery.

Caldwell Catalogue (C):

- Compiled by Sir Patrick Caldwell-Moore (1923–2012), an amateur astronomer and writer.
- Deep-sky objects.
- 109 items. Same as Messier List prior to M110 addition. Uses "C" as "M" was taken. Listed from most northerly.

New General Catalogue (NGC):

- Compiled in 1888 by John Dreyer (1852–1926), a Danish astronomer working in the British Isles.
- Deep-sky objects.
- 7,840 objects, leveraging the prior lists by William, Caroline, and John Herschel.

Henry Draper Catalogue (HD):

- Compiled primarily by American astronomer Annie Jump Cannon (1863–1941).
- Named for Henry Draper (1837–1882), an amateur astronomer from Virginia.
- Stars (locations and spectral classes).
- 359,083 stars in total.

Collinder Catalogue:

- Named for Swedish astronomer Per Collinder.
- Contains 471 open clusters.

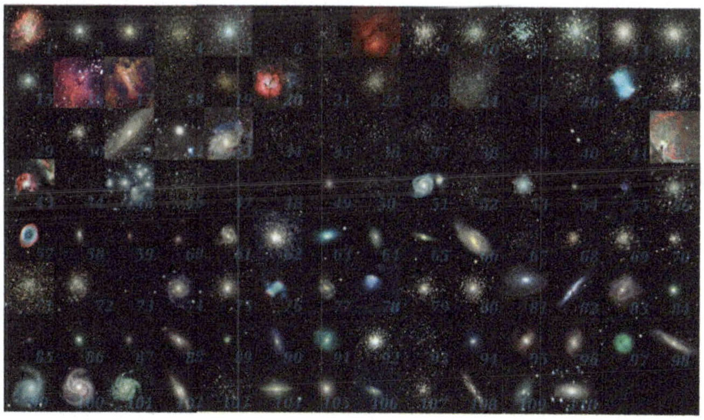

Images of all Messier objects, captured by Michael A. Phillips.

What You Can Expect to See

Objects like the Moon and planets can be viewed from the city, under streetlights, and even while using your phone. These objects are literally as bright as day and do not need fully dark skies, so they can be observed in light-polluted skies in the city or with the bright Moon present.

Individual stars, double stars, and open star clusters (see pages 20–21) can be viewed from the city or suburbs, but may be difficult to locate without the help of a Go-To telescope.

Galaxies, nebulae, and globular clusters, however, require a very specific set of conditions to maximize the quality of your view.

- It must be fully dark, that is, the Sun must be well below the horizon.
- The sky must be clear of clouds, haze, and smoke.
- The Moon must have either already set or not yet risen (New Moon evenings are ideal).
- You must be far from cities or towns.
- Your eyes must be adapted to see in the dark.

Adapting Your Eyes to the Dark

How do you adapt your eyes? You prepare your eyes to view deep-sky objects by avoiding looking at all sources of light for at least twenty or thirty minutes. This means avoiding looking toward porch lights, cell phones, and car headlights.

Some people on my YouTube channel have talked about using eye patches, but I've never actually seen anyone use this technique in real life.

Astronomers often use red lights to help preserve their night vision.

A Note on Color

Did you know that the human eye can only see in black and white in dim lighting?

Only when you use a camera do galaxies and nebulae show their color. Many objects imaged using professional telescopes aren't even in wavelengths the human eye can see. In this case, professional astronomers assign a color to that particular wavelength of light. This is called "false color" or "representative color."

This book is about what you can **see** through your telescope, not what a camera can capture. Amateur astronomers often call deep-sky targets "beautiful smudges," because without a camera, that is how many deep-sky objects appear.

All "telescope view" photos of galaxies and nebulae in this book are in grayscale to simulate how they look through the eyepiece. These images are examples of what you can reasonably expect to see under ideal conditions.

These images were taken with a variety of telescopes and show varying degrees of magnification. Your view may differ depending on the telescope and magnification you use.

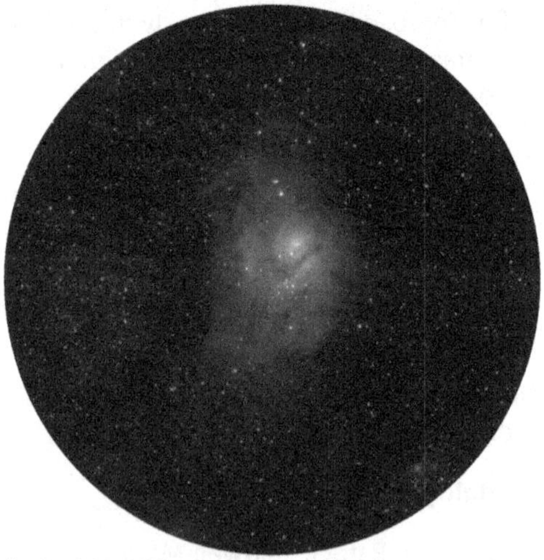

This nebula, M8, will appear pink in photographs but gray through a telescope.

Brightness

In astronomy, brightness is measured in a unit called "visual magnitude" or simply "magnitude," which quantifies the total light emitted by an object. This measure is loosely related to how challenging an object will be to see. The scale seems "backward," with brighter and easier-to-see objects indicated with lower (or even negative) numbers and dimmer objects with larger numbers.

However, this relationship is not always straightforward. Many of these objects are "extended," meaning they are not made up of "point sources," like stars. A large extended object, even if bright, may appear dim to the eye. For example, the galaxy M33 has a magnitude of 5.7 but is more challenging to see than the Dragonfly Cluster with magnitude 6.4. This is because the Dragonfly Cluster consists of point sources, concentrating its light into more compact areas, whereas the light from M33 is spread out over a larger area.

NGC 457 (left) appears far brighter than M33 (right) even though it is less "bright" in terms of magnitude. This is because the light from M33 is spread over a large area.

Different Views With Different Tools

As mentioned elsewhere in this book, the tools necessary for stargazing depend on what you're planning to see. Constellations and the Milky Way are best seen with the unaided eye.

Although binoculars show a much less magnified view, this doesn't necessarily mean that an object will look better in a telescope.

Large objects, such as the Andromeda Galaxy and Alpha Perseid Cluster, are too big for telescopes, and are best viewed through binoculars.

However, binoculars are not great for viewing the planets, and are unable to resolve most deep-sky objects as anything other than tiny smudges.

Telescopes, on the other hand, typically feature high apertures (which determines resolution) and long focal lengths (combined with the eyepiece to determine magnification).

Higher apertures allow the stargazer to distinguish fine details in deep-sky objects. Almost any telescope should be able to easily resolve the rings of Saturn and its largest moon, Titan, which would be quite challenging with binoculars.

The image on the left shows how the Andromeda Galaxy might look through a large telescope under ideal conditions. Note how we're really only seeing the core of the Galaxy here.

On the right, we have a view of the Andromeda Galaxy as might appear with binoculars under ideal conditions. In this case, the entire galaxy as well as its companions (dwarf galaxies M32 and M110) are fully visible.

"Seeing" and "Transparency"

"Seeing" refers to how sharply objects appear. Poor seeing is often caused by atmospheric turbulence or heat waves, similar to those above a hot road. Sometimes, during poor seeing, planets and stars can appear almost as if they're on fire! To estimate this, observe the brightest star visible through your telescope and rate the clarity of its image on a scale from 1 to 5, with 5 indicating the best conditions:

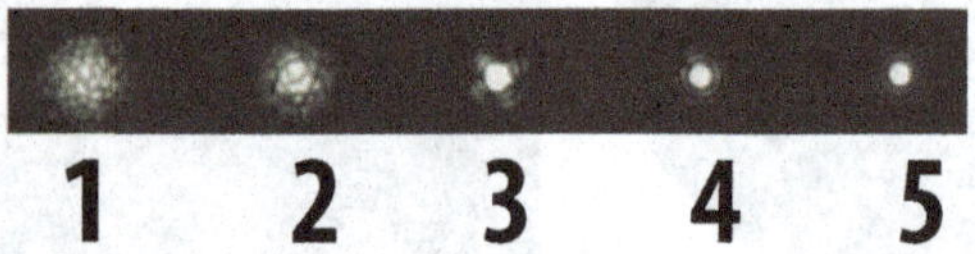

"Transparency" can be estimated by recording the magnitude of the dimmest star you can see with your eyes. I often use the stars in the Little Dipper to rank the transparency from 1 to 5 (with 5 being the best).

Estimates of seeing and transparency are typically required for most observing programs. In essence, if you are working on a certification or want your observations to be recognized by your astronomy club, you need to record the seeing and transparency for each night. You'll notice that the "Observing Log" for each target in this book has a row labeled Seeing and Transparency.

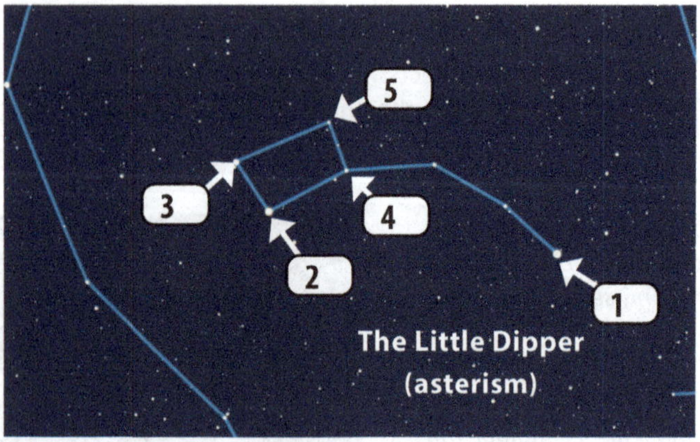

You can use this simple scale to estimate the transparency of the sky for completing your observing log. Simply record the number corresponding to the dimmest star you can see in Ursa Minor (the Little Dipper).

Stargazing Weather Applications

To estimate the seeing and transparency ahead of time, most astronomers use an app like *Astrospheric*. These apps can help you plan your observation by providing advanced notice of cloud cover and Moon phase as well as seeing and transparency.

Sky conditions are considered "ideal" when all three rows show dark blue during the period when you're planning to go stargazing. Of course, this assumes you're away from city lights and the Moon is below the horizon.

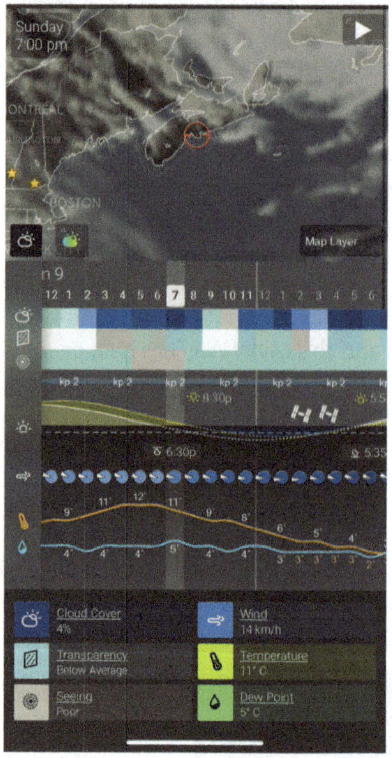

This free astronomy weather app is called Astrospheric. The blue boxes provide an hourly estimate of the cloud cover, seeing, and transparency.

As you can see from the highlighted column (7 PM), although the sky is relatively free of clouds, both the seeing and transparency are poor, most likely due to haze in the atmosphere.

Magnification

There are only a few situations, such as observing planets and the Moon, where you should use "high magnification" to zoom in on targets after you've found them at low magnification. This is because the planets and lunar craters cover a very small portion of sky, so it makes sense to "zoom in." Most targets, such as nebulae and star clusters, actually take up a fair bit of space in the eyepiece, so high magnification is not required.

The magnification of a telescope is how much larger objects appear compared to viewing without a telescope at all. It is the ratio of the focal length of the telescope to the focal length of the eyepiece. Note that a Barlow lens will multiply the magnification. Magnification can be calculated as follows:

Telescope focal length ÷ **Eyepiece focal length** × **Barlow strength**

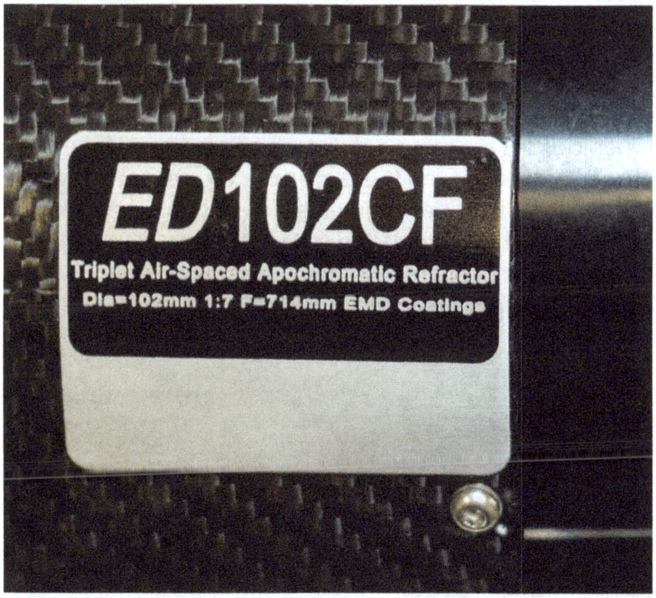

Telescopes often include a tiny plaque on the side of the optical tube indicating the telescope's aperture, focal ratio, and focal length. In this case, aperture = 102 mm, focal ratio = f/7, and focal length = 714 mm.

Resolution and Maximum Magnification

Magnification does not change the resolution of the telescope. Don't fall for misleading marketing: with beginner telescopes, the best views tend to be between 50x and 100x. Although an object may appear larger in the eyepiece, it won't necessarily show any more detail. For deep-sky objects like galaxies, magnification is primarily used to improve the contrast of an object, that is, how easy it is to distinguish from the background sky. The only way to improve resolution is to move to a larger (i.e., higher aperture) telescope.

The resolution of the telescope determines how detailed objects will appear. Therefore, **higher apertures provide access to higher magnifications.**

This leads to the maximum "useful" magnification of a telescope, which can be estimated as follows:

Telescope aperture (in mm) × **2** = **Maximum useful magnification**

Increasing magnification without purchasing additional eyepieces can be accomplished with a Barlow lens. However, these often make the telescope more difficult to use. More on Barlow lenses on page 35.

A typical 2× Barlow
Note: This goes between the telescope and the eyepiece; it does not function as an eyepiece by itself.

Types of Objects

The objects in this book fall into seven categories: stars, double stars, open clusters, globular clusters, nebulae, galaxies, and Solar System objects (i.e., planets, moons, and comets). With the right combination of telescope and eyepiece, all of these objects should fill you with a sense of awe and wonder.

However, these are not the only things you'll see in the sky. Satellites silently dominate the evening and morning sky, while airplanes with blinking lights and a gentle hum constantly pass overhead.

Stars & Asterisms

Most of the stars we see in the night sky are many times larger than the Sun. It's not that there are more giant stars (there aren't); it's just that they're so much brighter that large stars dominate our view. In any case, stars should appear as points of light, but brighter stars may appear larger simply due to their increased brightness.

Asterisms are chance alignments of stars that look interesting, so we assign names to them. It's like looking at clouds during the day and finding shapes. Large asterisms, such as the Teapot in Sagittarius, are often parts of constellations, but several others, like the 37 Cluster, can only be seen with a telescope, as the stars that form them are too dim for the unaided eye.

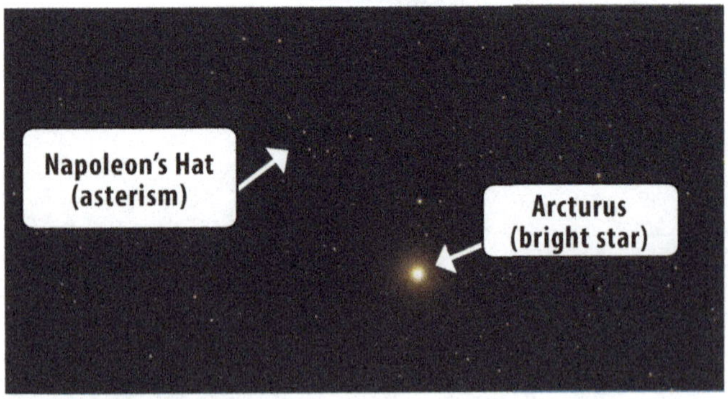

Arcturus (and Napoleon's Hat)

Double Stars

Many of the stars we see in the night sky have companions. In most cases, however, multi-star systems are so close together that their light blends together into a single source. Other times, it's just a chance alignment. For the purpose of this book, any two bright or interesting stars that appear to reside close to each other will be referred to as "double stars."

Zubenelgenubi (double star)

Open Clusters

Open clusters are groups of dozens or hundreds of stars that all formed around the same time. They reside in clusters simply because not enough time has passed for the stars to have diffused out into their galaxy. Over time, stars within these clusters become perturbed by outside gravitational forces, causing the clusters to dissipate.

Open cluster M11

Globular Clusters

Globular clusters are very different from open clusters. The main differences are that globular clusters reside outside the galaxy in a region called the Halo. The stars in these clusters are not perturbed by outside forces, making them almost like tiny elliptical galaxies.

Globular clusters contain hundreds of thousands, or even millions, of stars, and even a few black holes. They don't seem to contain star-forming regions, and therefore consist primarily of extremely old stars.

Assuming your skies are dark, the larger the aperture of your telescope, the more stars you'll see. The largest globular cluster visible from Earth is called Omega Centauri. However, it is not included as one of the targets in this book as it is only visible at southerly latitudes.

Galaxies

Galaxies are groups of millions to billions of stars held together by gravity. They typically come in two general categories: spiral galaxies and elliptical galaxies. Other than their shape, spiral galaxies tend to contain a lot of young, newly formed stars, with new stars forming all the time. Elliptical galaxies contain mostly old stars and lack the great star-forming regions of gas and dust.

Even in large telescopes, most galaxies appear as smudges. However, from truly dark skies, a spiral galaxy's spiral structure can start to take shape. The most impressive for me is M33, which resides near the Andromeda Galaxy.

Although elliptical galaxies tend to be larger, they're less common and less dense than spiral galaxies. Through a beginner's telescope under dark skies, the brightest elliptical galaxies will appear as small, round smudges.

Globular cluster M13 (Burke-Gaffney Observatory)

Spiral galaxies (the Leo Trio)

Elliptical galaxy M87

Nebulae

Nebulae are giant clouds of gas and dust. Some of these clouds are collapsing to form new stars, and are therefore called "star-forming nebulae." Others are the result of stars blowing off their outer layers at the end of their lives, which are referred to as "planetary nebulae" due to their often round, planet-like appearance. Some nebulae, like the famous Crab Nebula (M1), are supernova remnants, that is, what's left over after a giant star explodes.

Nebulae generally appear in varying shades of gray to the eye through a telescope regardless of size. However, add a camera, and all sorts of colors spring into view. Typically, pink and purple represent hydrogen atoms, re-emitting light from nearby stars (these are called emission nebulae). Bluish colors are caused by dust, and these are known as reflection nebulae.

The Trifid Nebula captured with a smart telescope Remember, to your eyes, these objects will appear gray.

The Dumbbell Nebula (M27)

Solar System Objects

Solar System objects include all the objects that reside within our Solar System. These include planets, dwarf planets, moons, asteroids, and comets. Viewing planets is where beginner's telescopes really shine. From the cloud belts and the Galilean moons of Jupiter to the rings of Saturn and the phases of Venus, viewing the planets through a telescope never gets old.

In recent years, there have been several comets just beyond visibility with the unaided eye. But, with the addition of binoculars or a telescope, they would pop into view, even from the suburbs. It's incredibly exciting to witness something you might never see again!

Comet C2023 A3 (Tsuchinshan-ATLAS)

Artificial Satellites

These days, artificial satellites (or simply "satellites") are generally considered more of a nuisance to astronomers than something to get excited about. However, several events are worth mentioning and will become more common as time passes. For example, the International Space Station, often the brightest object to pass overhead, recently got some competition—the Chinese space station Tiangong, which looks just as bright to me.

Rocket staging and booster burns are also becoming more frequent and can put on quite a show. These events can create colorful bursts of light and clouds, which would be a treat to chase across the sky with a small telescope.

You'll often see satellites "flair." This is when a reflective surface on the spacecraft reflects sunlight directly at your location. There are apps that can tell you when these flairs will occur, and they can be fun to photograph in long exposures.

Satellite flair (~20-second exposure).

SpaceX second stage burn

Apparent Size

Apparent size is how large an object appears to the observer. To the amateur astronomer, knowing an object's apparent size is important to understand how large it should appear in the telescope.

For the most part, an object's actual size is irrelevant. For example, the Moon covers about the same area of sky as the Triangulum Galaxy M33, despite the fact that M33 is about 100 trillion times wider. This is because M33 is about 100 trillion times more distant than the Moon.

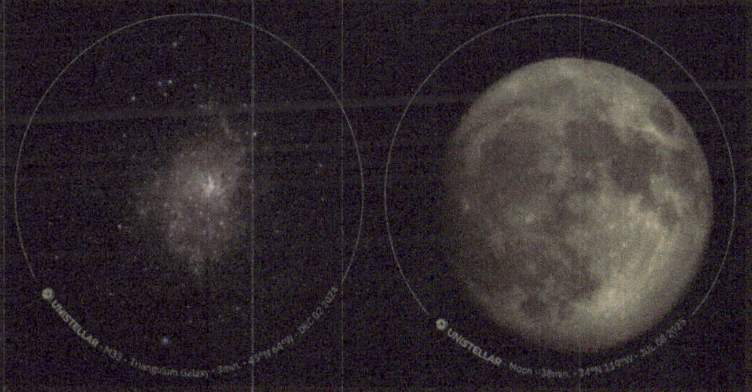

We measure an object's apparent size in terms of degrees or fractions of a degree. The sky is 90° from the horizon to zenith (straight up). The Sun and Moon are about half a degree in size. A degree is broken down into minutes of arc. In this case, the Sun and Moon are approximately 30 arcminutes wide.

An arcminute is broken down again into arcseconds, which are often used to measure the apparent size of planets or the separation between double stars. For example, Mars ranges in size from 3 to 25 arcseconds depending on its distance from Earth. Albireo, a double star, is separated by 34 arcseconds.

Degrees, arcminutes, and arcseconds are sometimes denoted by the symbols °, ′, and ″, respectively. For example, in the star maps in this book, a double star's separation will be listed with the star's name (e.g., Albireo 34″).

Choosing a Telescope

If you type "telescope" into a search engine, you'll see what appears to be hundreds of different types of telescopes. However, small beginner's telescopes fall into only a few categories.

Each of these telescopes are the right tool for *some* job, but you want to be sure you're getting the telescope that will resolve the particular targets you're interested in. If you're buying a telescope to view the Moon, then almost any telescope will do. However, some targets can *only* be seen with a camera, while some are best viewed with binoculars or with your unaided eyes!

When selecting a beginner's telescope you will want to consider:

- What you plan to observe
- Your budget
- How portable you want the telescope to be
- Your ability to reach dark skies
- Versatility with regard to upgrades and accessories.

Individual stars and open star clusters tend to be visible from the city or suburbs. However, finding their location can be challenging as fewer stars are visible from built-up areas. This is why some folks opt to use a Go-To telescope for stargazing activities in the city.

Some people also prefer a telescope that includes a quality finderscope for use in the city. Finderscopes with large apertures (such as 50 mm) cut through some of the light pollution, allowing you to identify the appropriate reference stars.

The main points that differentiate the telescopes into the categories that follow are: **focal length**, **aperture**, and **focal ratio**.

Specifications

Focal length: This is the distance from the primary lens or mirror to where the telescope brings light rays into focus. Placing a screen or camera sensor there would produce an image. In general, the higher the focal length, the longer the telescope's tube.

Aperture: This is the diameter of the primary lens or mirror. The primary lens is the big lens at the front of the telescope. For a Newtonian, the primary mirror is the large mirror at the back. In general, aperture is the most important quality of a telescope. Aperture determines the resolution of a telescope. Higher apertures grant access to higher magnifications without sacrificing fine details.

Focal ratio: This is the ratio between focal length and aperture. If this were a camera, it would be called the "f-stop." In stargazing, budget telescopes with higher focal ratios are less affected by **chromatic aberrations** (annoying halos around bright objects) than those with lower focal ratios.

In astrophotography, focal ratio determines the "speed" of a telescope—that is, how efficiently it can deliver light to each pixel on the camera sensor. A telescope with a lower focal ratio (e.g., f/6 or below) gathers light much faster than a slower telescope with a focal ratio of, say, f/10.

Magnification: This how much larger an object appears with than without the telescope. Typically, quality telescopes do not mention magnification as a selling point. This is because magnification is by far the least important quality of a telescope.

Magnification is governed by the eyepiece (more on this later), and telescope manufacturers typically include high-powered eyepieces and Barlow lenses to boost the magnification number mentioned on the box.

Anatomy of a Beginner's Telescope

Refractors, such as the one pictured, are the most popular type of beginner's telescope. The components outlined below are required to view objects in space with ease. Refractor telescopes sold without these parts can be extremely challenging (or impossible) to use for stargazing.

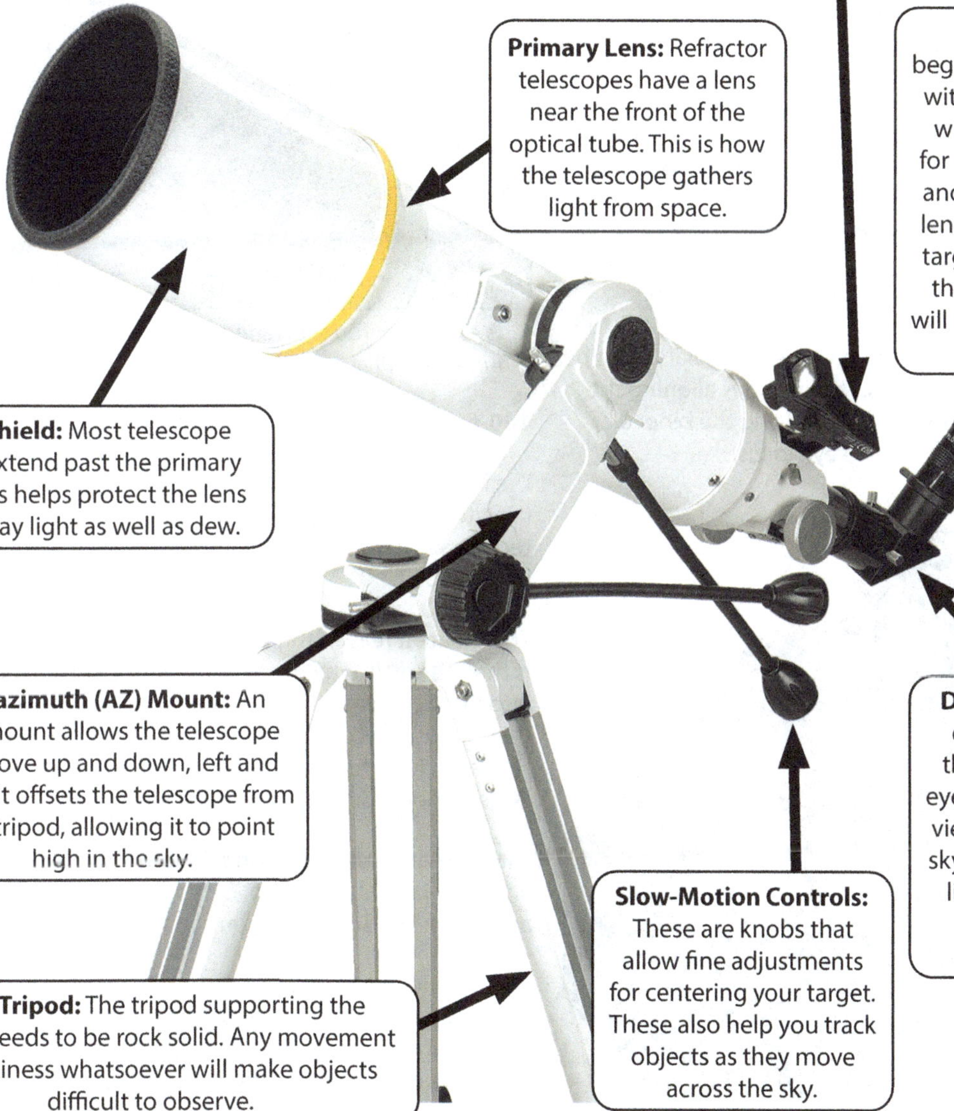

Finder: These days, most telescopes come with a red dot-style finder. These require 2032 batteries, and project a small red dot onto the sky. Finders must be aligned to the telescope prior to use.

Primary Lens: Refractor telescopes have a lens near the front of the optical tube. This is how the telescope gathers light from space.

Eyepiece: Typically beginner's telescopes come with two eyepieces—one with a long focal length for general observing, and another with a short focal length for "zooming in" on targets after you've found them. **Note:** a telescope will not function without an eyepiece!

Dew Shield: Most telescope tubes extend past the primary lens. This helps protect the lens from stray light as well as dew.

Altazimuth (AZ) Mount: An AZ mount allows the telescope to move up and down, left and right. It offsets the telescope from the tripod, allowing it to point high in the sky.

Diagonals: Diagonals direct the light from the telescope into the eyepiece allowing you to view objects high in the sky. Refractor telescopes like this should come with a 90° (not 45°) diagonal.

Slow-Motion Controls: These are knobs that allow fine adjustments for centering your target. These also help you track objects as they move across the sky.

Solid Tripod: The tripod supporting the mount needs to be rock solid. Any movement of shakiness whatsoever will make objects difficult to observe.

Types of Beginner's Telescopes

Binoculars

A telescope is not always the best tool for the job. For certain deep-sky objects (especially when you have dark skies), binoculars are by far the better choice. Even though most objects appear extremely small through the lens, binoculars offer an immersive view with both eyes and cover a wide portion of the sky.

"Zoom" binoculars offer another advantage in dark skies: changing the "zoom" seems to adjust the contrast between deep-sky objects and the background sky, making them appear to "pop out" from the background.

In addition to low magnification, the disadvantage of binoculars is that they're personal. Each observer needs to locate the object for themselves, and the binoculars must be adjusted for an individual's eyes. This includes aligning the binoculars along the central hinge, focusing the binoculars, and setting the diopter (the single eye focus).

Bushnell Legacy 10-22x50 zoom binoculars

Kid's telescope: 100-mm reflector with parabolic mirror (Zhumell Z100)

100-mm Reflectors: For Kids and Viewing the Moon

I've found the best kid's telescope to be a 100-mm reflector (Newtonian) telescope with a parabolic mirror. This is technically a Dobsonian design, which is extremely easy for kids to point.

The greatest reason this type of telescope is suitable for kids is that it's small, which makes it easy for kids to handle. It's also portable and easy to set up on a small table or rock. The magnification is quite low, which makes objects easy to find. However, for adults looking to go further in the hobby, a larger scope would be more fulfilling.

Examples include Celestron's "Moon Mission" telescopes, and the Sky-Watcher 100p. Other options include the Z100 and Orion Sky Scanner, if you can find them.

When looking for a portable telescope for kids, consider larger "small" Dobsonians (130 mm and 150 mm) such as the Sky-Watcher Heritage Series. Especially collapsible versions. These are often more portable than small refractors with large tripods.

Short-Tube Refractor

Short-tube refractors are the most common beginner's telescopes. My recommendation, if you're looking for a budget telescope, is to look for those with 102 (the aperture) in the name.

These telescopes are designed to be easy to use and offer wide fields of view. They are great for viewing large nebulae such as M8 and M42 from dark skies. If you'd like to get close-up views of the Moon and planets, you may need additional high powered eyepieces or even a **Barlow** lens, which sits between the eyepiece and the telescope to double or triple the magnification.

If you're thinking of purchasing a refractor, you need to be careful. Budget telescopes, especially those with apertures below 100 mm, often have design flaws that prevent them from being used effectively, which isn't any fun. Sometimes, it's because the mount is too shaky, the finderscope doesn't work, or the quality of the eyepiece is poor.

Long-Tube Refractor

Long-tube refractors are designed to provide crisp views of the Moon and planets on a budget. The long focal length helps eliminate the halos (called chromatic aberrations) that appear around bright objects when looking through a short-tube refractor.

Long-tube refractors have naturally high magnifications. This means that the Moon and planets can generally be viewed comfortably with moderate–focal-length eyepieces.

Long-tube beginner's telescopes with apertures smaller than 100 mm are typically best for your first views of the Moon or rings of Saturn, but are lacking when it comes to comfortably viewing deep-sky objects, like nebulae and galaxies.

Larger long-tube refractors, like the 1000-mm Explore FirstLight 102, provide excellent views of both planets and deep-sky objects from dark skies.

Short-tube refractor: Nat Geo 102 from Costco

Long-tube refactor: Explore FirstLight 102 (focal length: 1000 mm)

Newtonians

Newtonian telescopes on altazimuth (AZ) mounts are very easy for beginners to use. They often have larger apertures and wider fields of view than similarly priced short-tube refractors. Newtonian telescopes on equatorial (EQ) mounts are quite difficult to use because rotating the mount places the eyepiece in awkward locations.

Also, if the focal ratio is greater than f/6, the telescope likely has an extra lens to artificially increase its focal length. In a beginner's telescope, this is a sign of poor optical design. These are often nicknamed "Bird Jones" scopes.

In my experience, until you reach about 130 mm in aperture, budget Newtonians tend to offer poorer performance than budget refractors.

Newtonian: Explore FirstLight 114 (focal length: 400 mm). This is a great telescope for kids, but it has inferior optics when compared with a 102 mm refractor.

Dobsonians

Dobsonians are highly recommended by the astronomy community for their ease of use. This style of telescope typically offers the highest aperture for a given budget.

A Dobsonian is simply a Newtonian placed on a lazy Susan-type mount instead of a tripod. Many designs, like the Sky-Watcher Heritage series, are collapsible, making them extremely portable.

Dobsonian telescopes come in a few different styles. Tabletop Dobsonians are small, often collapsible, and are designed to be used on a picnic table. Traditional Dobsonians have a solid tube, and typically sit on the ground to be used while standing. Large Dobsonians often require a ladder to reach the eyepiece.

These telescopes are also the most popular telescopes for outreach, since people can learn to use them for themselves in just a few seconds.

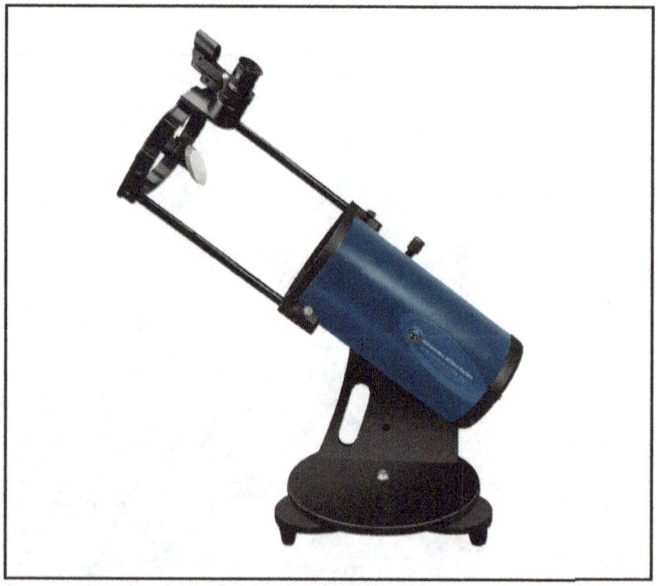

This 130-mm tabletop Dobsonian is based on the Sky-Watcher Heritage series. Proceeds from the sale of One Sky telescopes go to support STEM education programs.

CATs (SCTs and Maks):

Catadioptric telescopes, or "CATs," are at the higher end of the beginner's telescope market. These designs include Schmidt-Cassegrain telescopes (SCTs) and Maksutov (Mak) telescopes, which are compound designs that use specialized optics to compress a long focal length into a small package.

The main difference between a Mak and an SCT is that Maks use a thick correcting lens at the front and have a slightly longer focal length. The corrector plate of an SCT is lighter and flatter. The popular Celestron SE series of telescopes includes Maks for its smaller aperture models and SCTs for its larger aperture models.

CATs provide excellent performance and large apertures without the need for a long tube. When I'm asked which telescope "does it all," the SCT is my most common answer. Celestron SCTs can even be modified to take relatively wide-field photographs with an adapter called a HyperStar, which involves placing the camera in the place of the secondary mirror.

Astronomer Dave Chapman with a Celestron NexStar 8SE SCT telescope.

Smart Telescopes

"Budget" smart telescopes, such as the Seestar and Dwarf series, have become extremely popular. These "telescopes" are technically cameras—the user doesn't actually look at the stars but instead views the image on a phone or tablet. In fact, it's possible to photograph all of the deep sky objects in this book with a Smart Telescope.

In premium models, such as the Unistellar Odyssey Pro, the image can be viewed on an internal LCD screen. The view through this method is a live image created by stacking multiple exposures.

Smart telescopes are great for beginner astrophotographers as they are designed to track targets as they move across the sky and can stack a series of short exposures to reveal the structure of galaxies and nebulae in great detail. They're also great for capturing images of the Moon and Sun, but their focal lengths and apertures are generally too short for imaging planets.

These are a great option for travel, and the user experience gets better every year. I've scattered over 25 of my favorite smart telescope targets throughout this book, marking them with the figure shown on the right.

Using a smart telescope to capture an image of a distant lighthouse

Beginner's Telescope Mounts

It may surprise you to learn that any seasoned amateur astronomer will tell you that the mount is more important than the telescope! The mount, more than the telescope, will determine how "fun" a telescope is to use. A telescope mount should have the following characteristics.

- It should move smoothly and freely along its axes without jerky or jumpy movements
- It should allow for easy viewing when the telescope is pointed high in the sky
- It should stay where you put it without shaking
- It should allow you to track objects across the sky with a light touch on the optical tube or a gentle twist of the slow-motion controls

Altazimuth (AZ)

By far, the best mount for a *small* beginner's telescope is the altazimuth (AZ) mount. These mounts are characterized by an arm that extends from the top of the tripod, allowing the telescope to point high in the sky and rotate left and right over the tripod with ease.

AZ mounts often come with two slow-motion controls or a twist-to-lock panhandle. Even with slow-motion controls, the mount should still allow movement with a light touch on the optical tube.

Examples of beginner's telescopes on AZ mounts include the Celestron Omni 102 AZ and the National Geographic 102, and the Sky-Watcher StarTavel 102.

The Explore FirstLight series of telescopes typically includes a mount called the Twilight Nano. This mount does not have slow-motion controls, but as long as the telescope is well balanced, I've found this not to be much of a problem.

Premium AZ mounts include the Twilight 1 mount and the Sky-Watcher AZ5. Recently, brands like SVbony and some generic companies have begun producing heavier AZ mounts at lower price points.

But be careful. Many manufacturers place telescopes directly on top of camera tripods and call this an AZ mount. However, camera tripods do not allow the telescope to comfortably point high in the sky. They are not precise enough to stay put when you let go. In fact, I don't even refer to these as telescopes at all, as they more closely resemble spyglasses or spotting scopes.

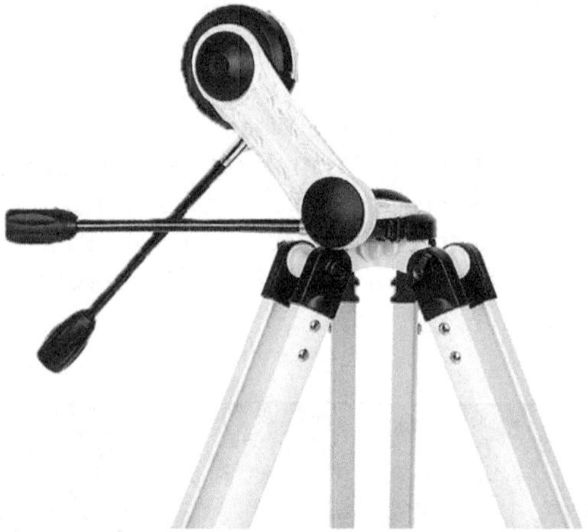

Lightweight AZ mount (from a Celestron Omni 102 AZ)

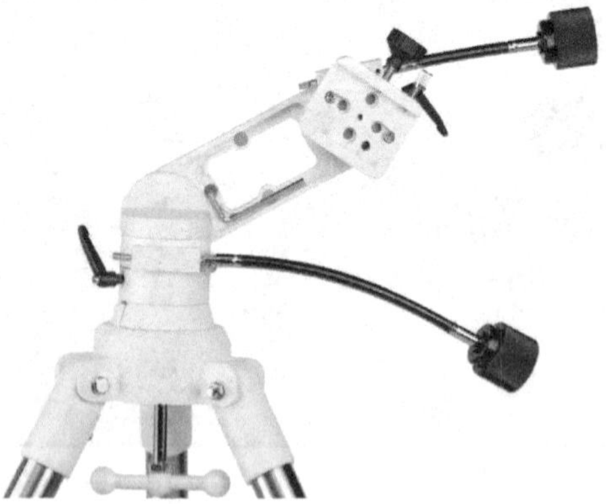

Heavy duty AZ mount (Explore Twilight Mount)

Go-To

Go-To mounts contain motors and a computer that move the telescope for you. These typically take about 15 minutes to set up. They require batteries or a power supply, and must be "aligned" to the sky.

The alignment process typically requires you to slew (move) the telescope to several bright stars, which teaches the computer where the mount is pointed. Depending on the model, you may need to identify these stars yourself, which can add to the challenge for beginners. A hand controller (or your phone) is then used to select targets that the telescope will slew toward on its own.

Go-To telescopes will also track objects as they move across they sky. However, Go-To AZ mounts are not designed for long-exposure astrophotography. Even though they track objects, the objects will rotate within the field of view. This effect, called "field rotation," is caused by the offset between the telescope's axis and Earth's rotational axis.

Popular beginner Go-To systems include the Celestron NexStar series, Sky-Watcher Virtuoso series, and Az-Gti.

Push-To

Push-To telescopes are similar to Go-To telescopes except that **you** are the telescope's motor. The telescope knows where it is pointed and where you want it to go. It then shows you on a screen (either on a hand controller or on your phone) which way to push the telescope.

Like a Go-To telescope, a Push-To system must be aligned to the sky. Celestron's StarSense system uses your phone's camera to perform the alignment. In this case, you must get a bright star lined up in the telescope itself and then select that same bright star on the image displayed on your phone.

Older systems that do not rely on a cell phone will require you to align the scope to a selection of multiple stars.

The most popular Push-To telescopes on the market today are Celestron's StarSense series, which I've found to be a LOT of fun. The Orion SkyQuest Dobsonian, one of the original Push-To systems, can often be found used and still works quite well.

Note that any telescope can be turned into a Push-To system using an app or website such as AstroHopper.

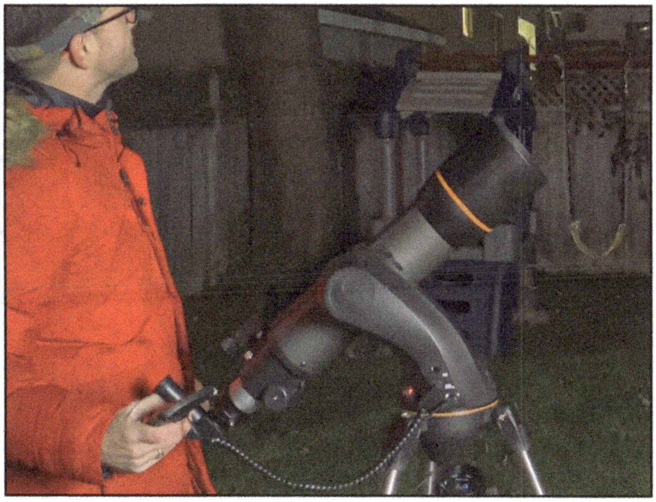

NextStar Go-To mount

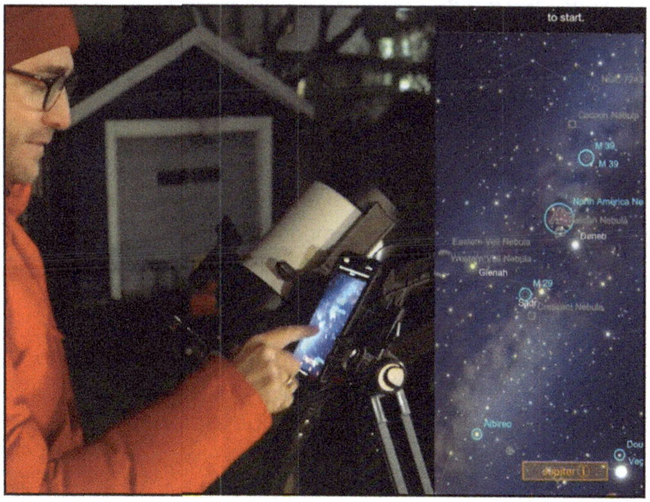

StarSense Push-To system

Rod-and-Yoke Mounts

Rod-and-yoke mounts are found on budget telescopes as low-cost alternatives to the traditional AZ mounts described earlier. While these mounts include a "stability bar" and "slow-motion knob" for the altitude axis, I've found that these additions do not provide much stability and they tend to be quite floppy when attempting to fixate on a target. To use these mounts, adjust the tension on the stability rod so that the telescope moves freely up and down but stays in place when you let go.

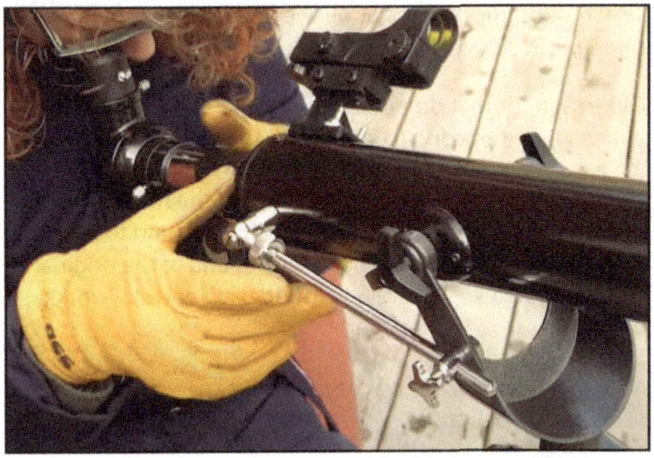

Rod and yoke mount

Camera Tripods

Many, many extreme budget scopes are sold on inexpensive camera tripods. Astronomers joke that these are so popular because people are buying them to spy on their neighbors, since they're clearly not designed to point at the sky.

If this is what you have to work with, see if the panhandle has a twist-lock feature. Often, twisting the handle counterclockwise will release the telescope, allowing it to point freely. After you've pointed it at a target, twist the handle clockwise to lock the telescope in place. When testing these telescopes, I have not had any luck tracking objects as they move across the sky.

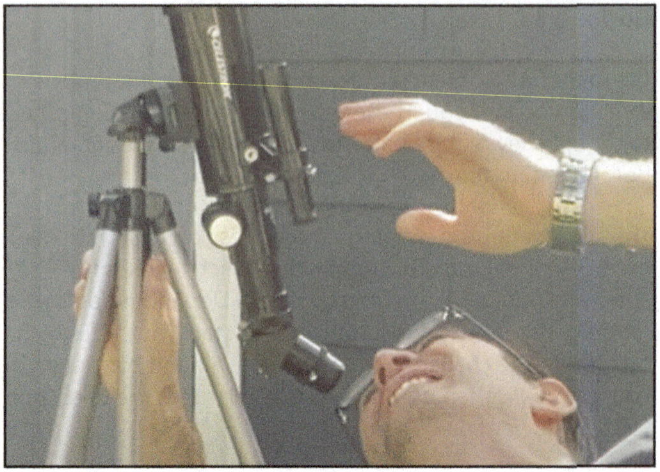

Telescope on a camera tripod (not fun)

Wedge Mounts

Wedges are designed to sit between an AZ mount and a tripod to convert the mount to an equatorial (EQ) mount so that it may track the sky along the RA axis. This is primarily done to convert a motorized AZ mount for use in astrophotography. Several mounts are capable of this conversion. I have converted my Sky-Watcher AZ-GTi to use a wedge (although it required a firmware update) and I have also seen some folks successfully convert a NexStar Evolution mount for astrophotography

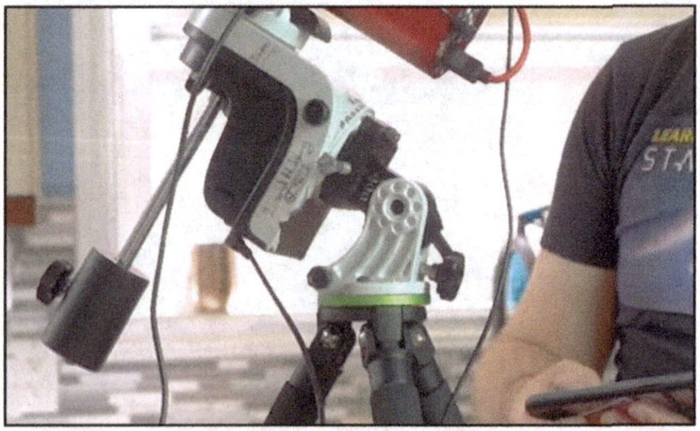

Wedge between AZ mount and tripod

EQ Mounts:

EQ mounts are designed so that you can track objects across the sky by turning a single knob, rotating the telescope's right ascension (RA) axis. I don't consider EQ mounts appropriate for beginners, as budget (sub-$500) versions of these mounts don't tend to function very well. Also, until the observer becomes familiar with how the sky moves, operating an EQ mount may be frustrating.

That said, high-quality, motorized EQ mounts are essential for astrophotography. These mounts work together with the camera to fix the field of view in place in the sky, making long exposures possible.

EQ mounts work by precisely counteracting Earth's rotation. The telescope's RA axis is aligned to the Celestial Pole. These mounts often include "setting circles" for referencing the telescope's position in RA, its position (east–west) around the **Celestial Pole**, and declination (Dec) the distance from the **Celestial Equator** (see page 14).

To use a manual EQ mount for visual observing:

1. Set the telescope in its "home" position (as shown). Lock the RA and Dec axes
2. Set the mount's altitude axis to your latitude
3. Rotate the mount's left–right (azimuth) axis so the telescope is pointing due north
4. Locate the Celestial Pole in the sky (in the northern hemisphere, Polaris, the North Star, is close enough)
5. Make fine adjustments to the mount's altitude and azimuth until the Celestial Pole is centered in the finder and eyepiece
6. Release the RA and Dec axes, and find your target (e.g., the Moon)
7. Rotate the optical tube within its cradle to position the finder and eyepiece comfortably
8. With the object centered in the telescope, re-lock the RA and Dec axes (the slow-motion controls will function with the axes locked)
9. Track the objects across the sky by slowly turning the RA slow-motion control

If this seems complicated (it is), I made a detailed video on this topic that can be found on the *LearnToStargaze* YouTube channel.

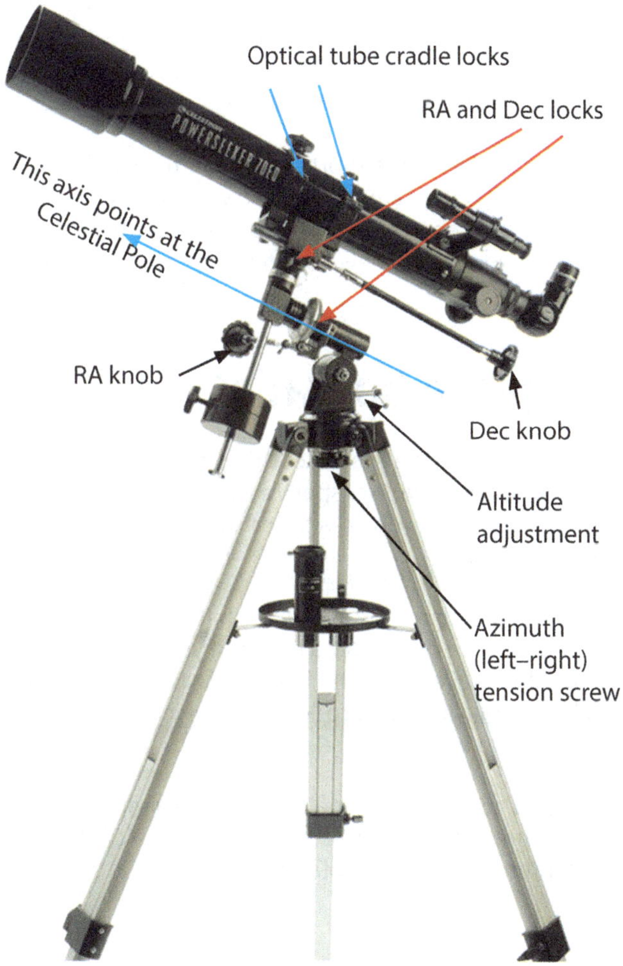

EQ mount in its "home" position

Choosing an Eyepiece

Beginner's telescopes generally come with two eyepieces. These are provided to optimize the views at two different levels of magnification. A telescope's magnification is the focal length of the telescope divided by the focal length of the eyepiece. In other words, the shorter the eyepiece focal length, the higher the magnification.

The focal length for one of these eyepieces is usually around 25 mm. This is the general-purpose eyepiece that you'll use most of the time. You can use this eyepiece for aligning the finder and for locating targets. In fact, for deep-sky objects like galaxies, nebulae, and star clusters, this may be the only eyepiece you'll use.

The eyepiece with the shorter focal length, usually 10 mm, is for zooming in on targets after you've found them with the higher–focal-length eyepiece. For deep-sky objects you'll usually find that the view is better at lower magnification.

Premium vs. standard eyepieces of the same focal length

Eyepiece Upgrades

Upgrading from the eyepieces that came with your telescope can provide crisper views and wider fields of view. Expect premium eyepieces to cost about as much as the telescope itself.

Tele Vue eyepieces are generally regarded as the "best" and highest-quality eyepieces on the market.

The focal lengths of the eyepieces that came with your telescope were chosen for a reason. If you are planning to upgrade the eyepieces, the best advice is to simply upgrade to higher-quality eyepieces of the same focal lengths as the ones that came with the telescope.

That said, for viewing planets, if you have a telescope with a focal length of less than 1000 mm, I've had a lot of success with premium 5-mm eyepieces.

Baader Hyperion eyepieces are a massive upgrade but at a much lower cost than Tele Vue.

Barlow Lenses

A Barlow is not an eyepiece and should be used sparingly. And in most cases, not at all. Some telescope retailers don't even sell them as they often cause so much trouble for beginners.

The Barlows that come with beginner's telescopes are typically included only to increase the magnification listed on the box. In my experience, their quality is often so poor that they barely function at all. Barlows often push the telescope past its maximum useful magnification (which can be estimated by taking twice the aperture in millimetres).

That said, a decent quality 1.25-inch Barlow can be purchased for about $30. If you want to experiment with Barlows, they can provide closeup views of the planets. In large-aperture telescopes, you may find that a Barlow increases the contrast in the image.

Generic 2× Barlow

Zoom Eyepieces

As an alternative to inserting a Barlow lens, consider using a zoom eyepiece to adjust the magnification of the telescope with a simple twist. For beginners, a zoom eyepiece may be more convenient than swapping eyepieces and inserting a Barlow. The downside is that the field of view is often narrower compared to premium single-focal-length eyepieces. Zoom eyepieces allow you to find the optimum magnification for your target. To use them for deep-sky objects, slowly reduce the focal length until the contrast on the object is maximized. Note that you may need to adjust the telescope's focus with each adjustment to the magnification.

Celestron zoom eyepiece

When to Use a Filter

Filters are added between the eyepiece and the focusing assembly or diagonal. These typically come in two sizes, 1.25 inches for a 1.25-inch eyepiece and 2 inches for a 2-inch eyepiece. Filters tend to be used sparingly, but are used to enhance the view of very specific objects.

Moon filters, or polarizing filters, are used primarily for the Moon. Light becomes polarized when it's reflected. Filtering this polarized light not only decreases the brightness, but also increases the contrast.

Color filters are used to increase the surface details on the planets. They work by blocking certain wavelengths of light. A lot of this is subjective and takes a bit of experimentation, but here are some things that have worked for me.

A blue filter increases the contrast on Jupiter, allowing increased contrast when viewing the clouds on its surface and attempting to pick out its Great Red Spot.

Most of the time, Mars will look like a bright red star. But at opposition, when it's closest to Earth, moderate magnifications will bring the disk of Mars into view, and it's worth experimenting with the yellow filter to try to bring out the variations in surface colors, and even to attempt to identify the polar ice caps.

Generic moon filter

Using a Beginner's Telescope

Focusing a Telescope

In order for you to see anything through your telescope, an eyepiece must be inserted and the telescope must be brought into focus. In a refractor or catadioptric (e.g., SCT), the eyepiece is placed in the diagonal. In a Newtonian, the eyepiece is placed directly in the focusing assembly.

A telescope must be focused each time you set it up, every time you change eyepieces, and every time you change users, as everyone's eyes are different. It's easiest to practice focusing during the day.

Step 1: Point the telescope at a distant, bright object.

Step 2: Turn the focusing knob until objects appear crisp and sharp. At night, stars should appear as tiny pinpoints of light.

Note: Some telescopes have "dual-speed" focusers with an outer knob for rough focus and an inner knob for fine focus.

If your telescope is set up and focused correctly, you should see a circular image. The stars should appear as tiny speckled points of light.

Aligning the Finder

To function effectively, both the finder and the telescope must be pointed at precisely the same spot. In order to make sure there's not a slight difference, you must align the finder to the telescope. This is easiest to do during the day.

Step 1: Add a low-powered eyepiece and center the telescope at a distant object. Examples include the top of a flagpole or a bright star.

Step 2: Without moving the telescope, move your eyes over to the finder. Use the adjustment screws on the finder to center it at the target.

Step 3: Move back and forth between the finder and the eyepiece to make sure that both are pointed at precisely the same spot.

Aligning the finder on a distant object

Check out the Learn To Stargaze YouTube channel for a very detailed video called "How to Use a Beginner Telescope."

Collimation

If you have a Newtonian telescope, you must know how to check the collimation, or the alignment of the mirrors. While smaller Newtonians require less frequent collimation, they usually require alignment before their first use. The easiest way to check a telescope's collimation is to point at a bright star and bring the telescope slightly out of focus. If the "donut" shape is not symmetrical, the telescope requires collimation.

Collimating a Newtonian telescope can be accomplished in two ways.

Method #1: Visually

Step 1: Remove the eyepiece and insert a collimation cap into the focuser. A collimation cap is a plug with a hole in the center. **Note:** A "Cheshire eyepiece" can also be used here and the concept is basically the same, except you'll have additional crosshairs to help center align the two mirrors.

Step 2: Observe the reflection of the primary mirror in your field of view. The clips holding the primary mirror in place should be symmetrical within your field of view. If not, adjust the three screws on the secondary mirror until the view is centered.

Step 3: Observe the length of the spider arms in the reflected image. These should appear of equal length. If not, loosen the three locking screws on the back of the telescope and adjust the three adjustment knobs until the spider arms are of equal length. Carefully retighten the locking screws.

Method #2: Laser Collimator

Step 1: Remove the eyepiece and place the laser collimator in the focuser.

Step 2: Adjust the screws on the secondary mirror assembly until the beam is centered on the primary mirror.

Step 3: Loosen the three locking screws on the back of the telescope and adjust the three adjustment knobs until the laser beam is centered in the collimator's screen.

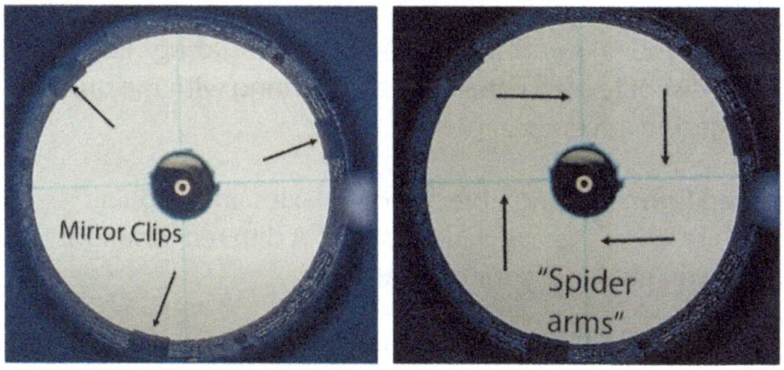

A correctly collimated Newtonian
View through the focuser without an eyepiece

Note: For instructions on how to get the most out of a smart telescope, see the astrophotography chapter near the end of this book.

Telescope Care

A telescope should function for years with minimal care. But there are a few things you can do to keep your equipment in optimal condition.

1. Telescopes often get wet with dew after a night of use. Make sure the telescope is fully dry before packing it up.

2. If a telescope's primary lens gets dirty, use an approved lens cleaning kit from a camera store to clean it. A little cleaning bush (shaped like a paintbrush) is helpful for cleaning difficult-to-remove dust from eyepieces. Note: A Newtonian telescope's mirrors do not need cleaning unless they become extremely dirty. I recommend consulting a YouTube tutorial for the cleaning process.

3. To prevent the telescope from gathering dust, cover when not in use. Always place the cap over the primary optics and put away the eyepieces in their designated holder.

The Astronomer's Toolkit

Other than your telescope and eyepieces, what else should you take with you while going stargazing? Here is a checklist of few essential items that I bring with me on any stargazing adventure.

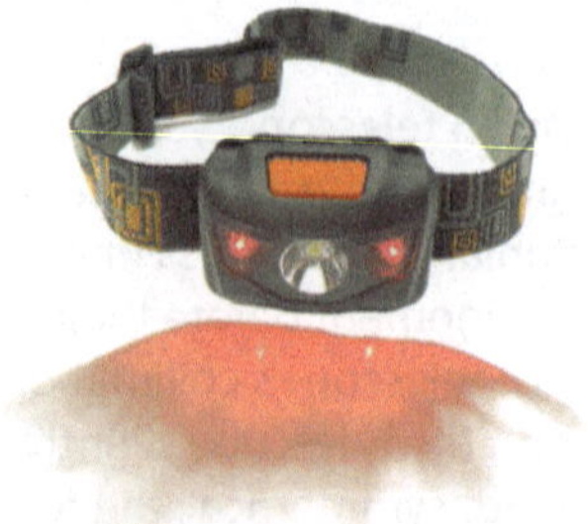

Red headlamps can be found online for about $10. Just make sure not to shine yours into anyone's eyes while they are trying to stargaze.

A red light: Viewing objects like galaxies and nebulae requires your eyes be dark-adapted. A dim red flashlight will help you maintain your night vision.

A stargazing guidebook: Stargazing guidebooks organized by season (like this one) provide a selection of ideal targets to simplify your stargazing experience.

A pencil: Recording what you see makes you an astronomer. That's why each page of this book contains an observing log and a sketching circle. Additional circles are provided in the back of the book.

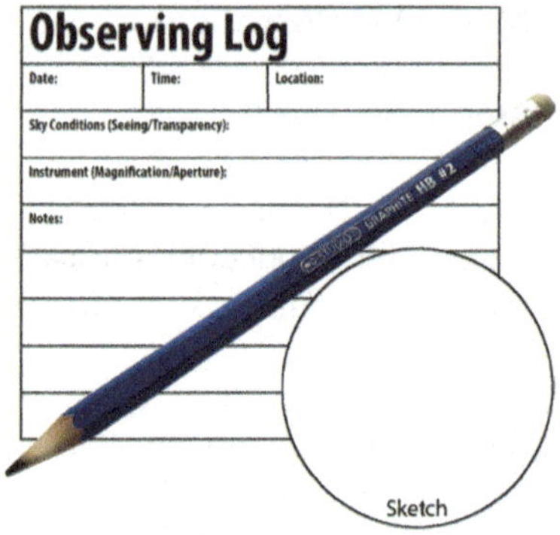

Binoculars: In additional to your telescope, binoculars are helpful for locating large, bright targets like clusters and galaxies or even comets. Depending on how dark your skies are, these may be difficult to see with the unaided eye, but they'll pop right out through binoculars.

10x50 binoculars seem to provide the best stargazing views.

Chairs: Binoculars and refractor telescopes are far easier to use while sitting down.

I use an Adirondack chair for both binocular stargazing and observing with telescopes. However, some astronomers prefer specialized adjustable stargazing chairs, like the one on the right. I have one of these in my basement, but I've never used it.

Jacket and hat: Even if I'm stargazing in summer or in California, I usually get cold just standing around. Warm clothes are a must on any stargazing adventure.

Batteries for your finder: Most finders these days take CR2032 batteries. It's best to have these on hand. If your finder battery dies and you don't have a replacement, stargazing with a telescope becomes a whole lot less enjoyable.

Bug spray: Trust me on this one.

Optional Accessories

I've found that many beginners want to get a lot of toys to go along with their telescopes at the time of purchase. However, my advice is to learn to use what came with your telescope, and only purchase accessories to solve specific problems that arise along the way.

If you've got a new telescope, the only accessories you really want are this book and a red flashlight to read it with.

That said, here are a few items you might consider adding to your stargazing wish list:

- **Dew heater:** If you find that your telescope gets covered in dew because you live in a moist climate, then this just might be your first accessory.
- **Extra red flashlights and red headlamps:** You want everyone in your group to be using red lights. That way, no one person ruins the stargazing experience by exposing everyone to white light.
- **Upgraded finder:** Upgrading from a finderscope to a red dot or from a red dot to a bullseye (reflex) finder can make objects easier to find.
- **Zoom eyepiece:** Zoom eyepieces work by changing the focal length of the eyepiece itself. These can be a lot of fun and help you reach the optimal magnification (and thus contrast) for a given target.
- **Cell phone adapter:** Many beginner stargazers want to dabble in astrophotography in order to share the view through their telescope. Attempting to capture images through your telescope with your phone is often the first step into the complex world of astrophotography.
- **Filter:** While I rarely use a filter when stargazing, seasoned observers commonly use these to take their views to the next level.

Generic USB dew heater

Celestron zoom eyepiece

Bullseye Reflex Finder

Celestron NexYZ cell phone adapter

UHC filters increase the details visible on certain nebulae.

Using This Book

The following sections contain all the information required to locate and observe the most popular targets. The book is organized by season, making it easy to pick targets on any night of the year. These are the meteorological seasons, meaning that summer begins on June 1, autumn begins September 1, winter begins December 1, and spring begins March 1. Note that the summer targets are visible late into the autumn due to the shortening days.

Although this book aims for one showcase target per page, most pages contain nearby objects that you may want to observe since you're in the area anyway.

Target Name: The target's name is listed at the top of each page. Often, a designation is included, which is helpful if you need to find the target in stargazing software.

Observing Log: You are encouraged to record your observations. These logs can be used by your local astronomy club and go toward awards and certificates.

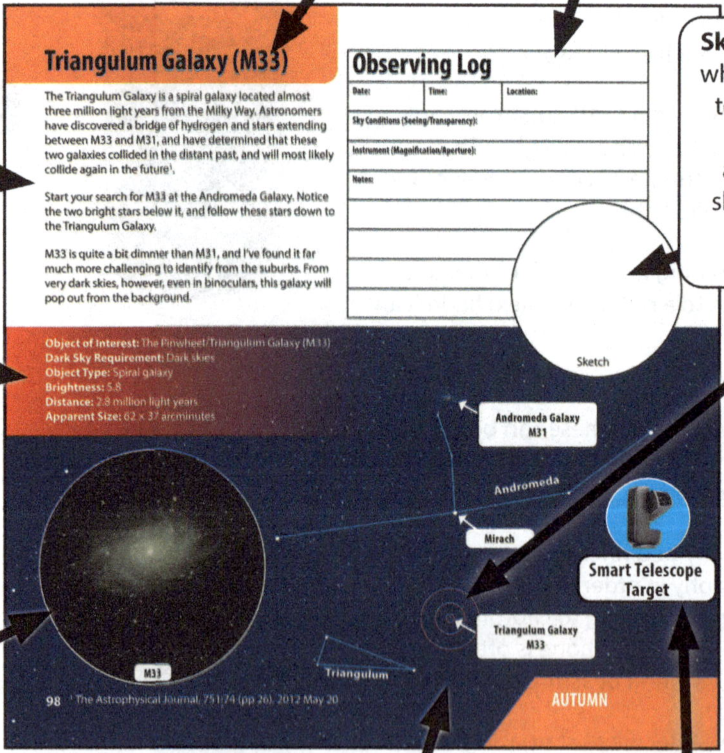

Text: A detailed description of the target and why it is worth observing. Starhopping instructions are also included here.

Fact Box: Displays facts about the object, including brightness, distance, and apparent size.

Telescope View: An approximation of how an object will appear under ideal conditions at the eyepiece. Simulates a level of magnification appropriate for the particular target.

Sketching Circle: Sketching what you see is the best way to record your experience at the eyepiece. Some astronomy clubs require sketches to be included as you apply for observing certificates.

Telrad/Bullseye Ring: I've found it helpful to include a bullseye ring for certain targets. If you're using a bulleye finder, aligning these rings as shown in the star map will help guide you to the target. These rings also correspond to the approximate field of view of 0.5° (inner ring), 2° (middle ring), and 4° (outer ring).

Star Map: These star maps show **only** the constellations or star patterns required to help you find your target.

Smart Telescope: My favorite targets to capture with a Smart Telescope.

How to Use Stargazing Software

Stargazing software is extremely powerful and doesn't need the internet to function. Simple, and often free, mobile apps can display an accurate image of the stars and planets for virtually any time in the past or future.

Stargazing software is useful for checking the phase of the Moon, for locating planets, and learning the positions of constellations. It's also helpful, when locating newly discovered comets, since the comet's orbit can be downloaded into the software (a detailed description on how to do this can be found on page 57).

As useful as stargazing software is, it has its limitations. For beginner stargazers viewing deep-sky objects like galaxies and nebulae, a stargazing guidebook (like this one) is FAR more useful. This is because stargazing software generally oversells what you'll realistically be able to see. Even the "tonight's best" features are often overly optimistic.

Stellarium

The two most popular stargazing programs are SkySafari (paid) and Stellarium (free). In this book, I'll describe Stellarium because it's my platform of choice for beginners.

Getting Started

Step 1: The first thing you'll want to do when you run stargazing software is confirm your location and time zone. In Stellarium, that's this button: 

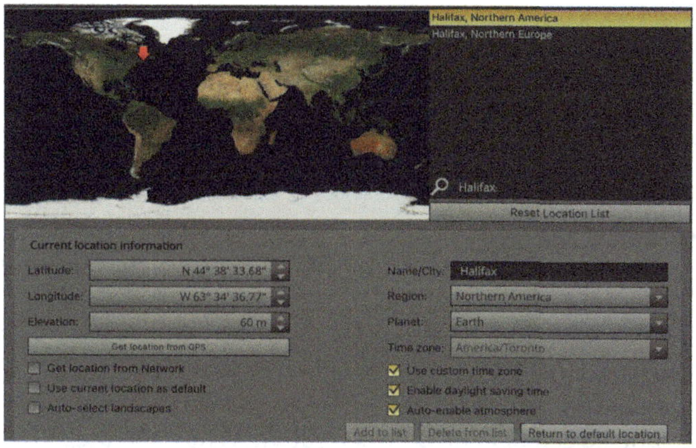

Step 2: Use the "Time and Date" window to set the time to the exact hour in which you plan on observing. 

Step 3: Turn on the constellation lines, constellation names, and planets. Turn off meteor showers, deep-sky objects, and satellites. The Stellarium menu bar below shows this configuration.

Solar System Objects

I think the main reason most people pick up a beginner's telescope in the first place is to observe the Moon and see the rings of Saturn. This section will show you how to do just that.

Most of the items in this section do not require dark skies or for you to travel out of the city. The Moon and planets can look just as good from your backyard as from the top of a mountain in Hawaii.

The only real exception here is a total solar eclipse, which in most cases would require you to travel to a specific location for the full effect.

Several of the events mentioned in this section, such as conjunctions, may only be visible at certain times of the night—a conjunction between Saturn, Venus, and the Moon, for example, may only be visible in the late morning or early evening, as Venus never strays far from the Sun. It's important to understand how to validate the time you can make these observations ahead of time using astronomy software (see section on this topic on the previous page).

Conjunctions and Occultations

Conjunctions make a good excuse to gather people together to appreciate the night sky. For the purposes of stargazing, a conjunction is any time objects appear together in the night sky. For example, over the course of its 27-day orbit, the Moon will have a conjunction with every planet in the Solar System as well as some bright stars. That doesn't mean all of these conjunctions will be noteworthy, but some will.

Sometimes you get a grouping of three objects together, such as the Moon, Jupiter, and Venus. When that happens, it's worth gathering your friends and having a bit of a stargazing party.

Lunar occultations occur when the planets (or bright stars) appear to pass behind the Moon. These can be a bit more challenging to observe as the observations are often brief, lasting only a few hours, and they are only visible from certain locations. For example, while an occultation of Jupiter by the Moon may be visible from South America, stargazers in North America may only witness a conjunction. Jupiter would simply pass near the Moon without going behind it.

The Moon about to occult the star Al Kaff (Gamma Ceti) on March 20, 2018

Event: Planetary conjunction
Dark Sky Requirement: City-viewable

The Moon (Full Moon)

If you're new to stargazing with a telescope, then the Moon is where you should start. The Moon is easy to find and will allow you to learn about the capabilities of your telescope, such as what happens when you change eyepieces. The Moon also changes from night to night as it goes through its phases.

When looking at the Moon through a telescope or binoculars, you'll first notice a number of large dark gray patches. These are the lunar seas or "maria" (pronounced mah-ree-ah). The light areas are known as the "highlands." You'll also see plenty of craters and mountain ranges.

Around many of the craters, you'll see ejecta, debris spewed up from the original impact. From craters like Tycho, you'll see rays of bright ejecta radiating great distances across the surface of the Moon.

The Moon is visible for about half the month in the evening sky. This is because the Moon goes through its phases every 29.5 days. I am often surprised when, on Moonless nights, some folks seem to think that we can see the Moon with the use of a telescope. Just to clarify: if you can't see the Moon without a telescope, you won't see it with one.

"Ray" Craters visible during the Full Moon.

Observing Log

Date:	Time:	Location:
Sky Conditions (Seeing/Transparency):		
Instrument (Magnification/Aperture):		
Notes:		

Sketch

Where is the flag?

People often ask me why we can't see the landers or the flag in a telescope. The Moon is about 380,000 kilometres away and about 3,500 kilometres across. Say you have a 16-megapixel camera and you take an image of the Moon that covers the sensor, which is about 5,000 pixels wide. That's 700 metres per pixel!

However, telescopes are also "aperture limited," meaning that the smallest detail it is possible to see is based on the aperture. If you wanted to see an object that is only one metre across, you'd need a telescope with an aperture of more than 100 metres. This is based on a law of physics called the Rayleigh Criterion.

Object of Interest: Full Moon
Dark Sky Requirement: City-viewable
Brightness: –12.7 (only when full)
Distance: 384,400 km (on average)
Apparent Size: ~30 arcminutes

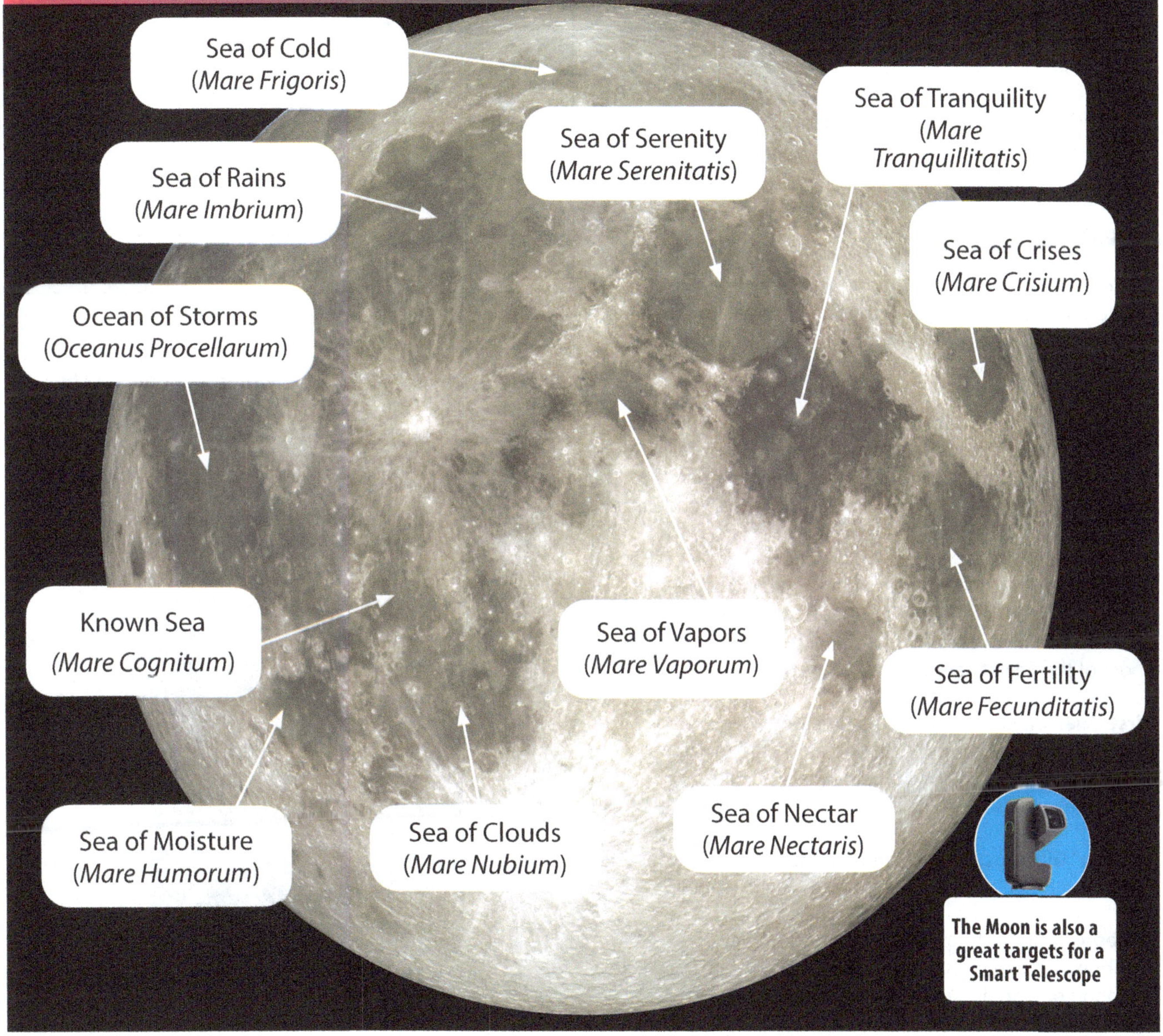

The Moon is also a great targets for a Smart Telescope

Lunar Phases and Major Features

As the Moon orbits around Earth, it passes through "phases" that repeat every 29.5 days. The Moon is visible in the evening sky right after sunset, only from the young crescent phase (a few days after the New Moon) until the Full Moon about 12 days later. After the Full Moon, it rises later and later each night. If you observe the Moon in early morning, some features will look a little different because of the changes in illumination. For the purpose of this book, I assume you're observing in the evening and so will ignore the phases after the Full Moon.

The craters that will appear most prominent are those located near the "terminator," the line between the daytime and nighttime sides of the Moon. Here are a few features to look for during these phases:

Young Moon: When the Moon is just a thin sliver, look for four large craters running along the terminator (the line between lunar night and day). These are nicknamed the "Gang of Four" (from top to bottom: Langrenus, Vendelinus, Petavius, and Furnerius).

Crescent Moon: Now there are many craters visible; the most easily identifiable are Cleomedes on the shore of the Sea of Crises, and Posidonius, a smooth-floored crater on the shore of the Sea of Serenity.

First Quarter: Look for the Apennine Mountains. Many large and bright craters can be observed along the terminator. Sunlit features like the Dumbbell Craters (Cyrillus and Catharina) are also visible.

"Young" Gibbous: Shortly after the first quarter, prominent craters such as Plato, Copernicus, and Clavius appear.

Gibbous: During this phase, Sinus Iridum ("Bay of Rainbows") is clearly visible, as is Gassendi, which looks like a diamond ring at higher magnifications.

Observing Log

Date:	Time:	Location:
Sky Conditions (Seeing/Transparency):		
Instrument (Magnification/Aperture):		
Notes:		
Use sketching circles located in rear of book.		

"Nearly" Full: When the Moon is almost full, look for the small but bright crater Aristarchus and two giant craters, Grimaldi and Shickard.

After Full Moon, the Moon rises later and later in the evening and into the morning. These phases, when it goes from Full Moon to New Moon, are called the waning phases. As I mainly stargaze in the hours after sunset (and not in the morning), I seldom encounter these phases.

There so many amazing features to see on the Moon with a telescope that I just had to write an entire book about it!

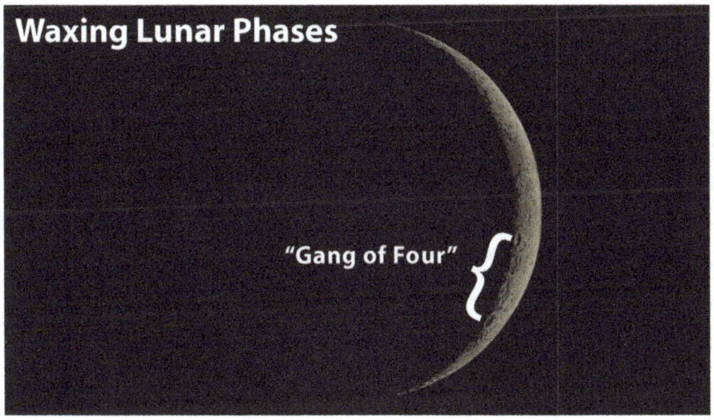

Young Moon. During this phase, the Moon is very low in the sky and sets soon after sunset.

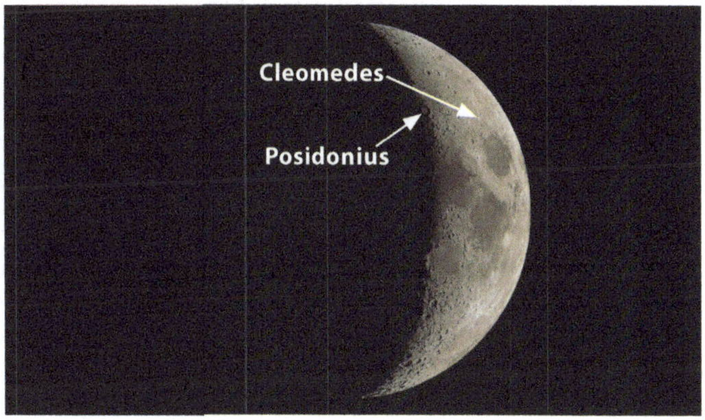

Crescent. During this phase, the Moon sets within a few hours after sunset.

First Quarter. During this phase, the Moon rises around noon (yes, you can see it during the day) and sets at midnight.

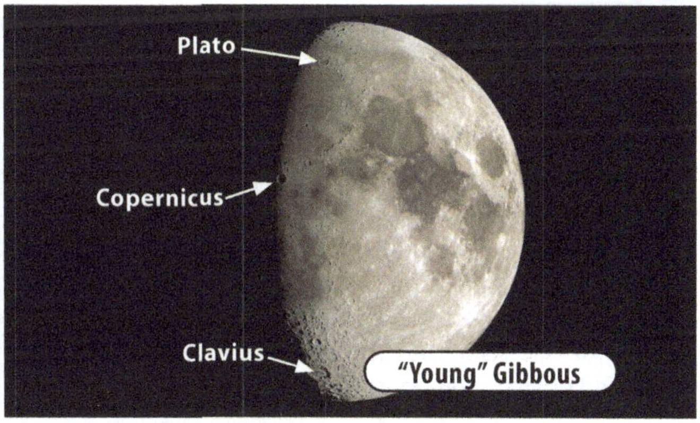

(Young) Gibbous. After First Quarter, the Moon rises in the afternoon and stays in the sky late into the night.

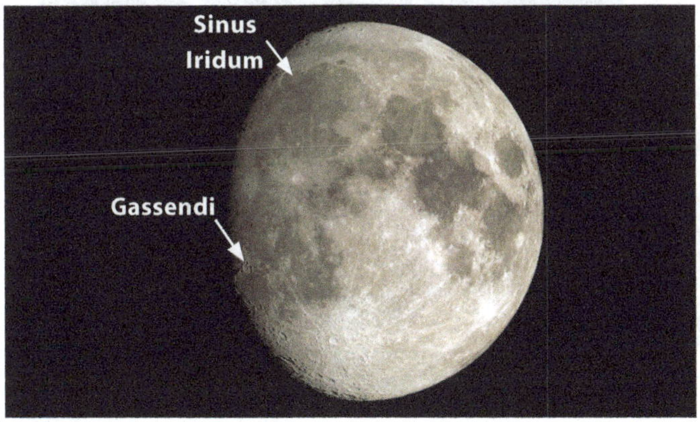

The Gibbous Moon is nearly full, and in the sky most of the night. Most of the lunar seas are visible.

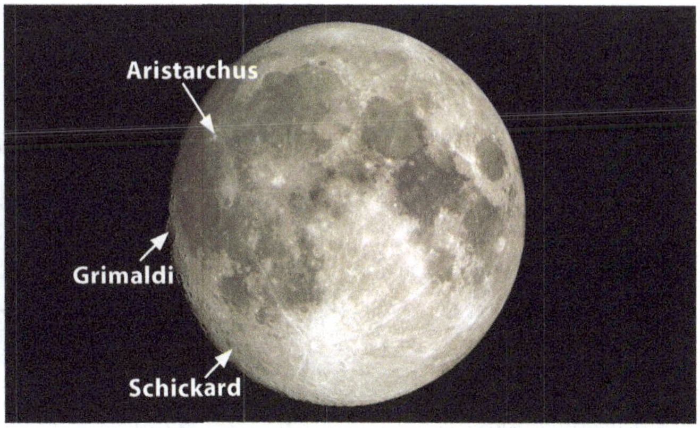

The (nearly) Full Moon is in the sky most of the night. All the lunar seas are now visible.

Jupiter and its Moons

Along with Saturn and the Moon, Jupiter is by far one of the most impressive objects to see with a telescope. On a night with steady air, even the smallest telescopes can distinguish its two main cloud belts and four "Galilean" moons. These four moons were discovered by Galileo, who was the first person to train a telescope on the planet Jupiter.

Moderately sized telescopes can even be used to identify the Great Red Spot, which will appear as a brownish perturbation in one of the cloud belts.

The moons Io, Europa, and Ganymede are found close to Jupiter, while Callisto is generally observed a bit farther away. But all four can be contained in a single field of view through the eyepiece.

Astronomy magazines often include a helical diagram showing the changes in position of the Galilean moons over time. Their locations are highly predictable and can be read like a clock. In fact, before the age of instantaneous communication, these moons were often used to synchronize clocks so that astronomical observations could be carried out simultaneously across the globe.

Jupiter has ~95 moons, but only these four are visible in a backyard telescope. The other moons are much smaller, and most orbit much farther from the planet. The fifth largest moon is Amalthea, an irregularly shaped rock only 250 kilometres at its widest point. It was discovered visually by astronomer E. E. Barnard in 1892 using a telescope with a 36-inch aperture.

The Galilean moons as captured by NASA spacecraft

Observing Log

Date:	Time:	Location:
Sky Conditions (Seeing/Transparency):		
Instrument (Magnification/Aperture):		
Notes:		

Sketch

Jupiter and the Galilean moons as they would appear through a telescope

Object of Interest: Jupiter (and the Galilean moons)
Dark Sky Requirement: City-viewable
Object Type: Planet (gas giant)
Brightness: –2.9 (brightest) –1.7 (dimmest)
Distance: 778 million km (from the Sun)
Apparent Size: 30 to 50 arcseconds

Jupiter imaged by the Hubble Space Telescope

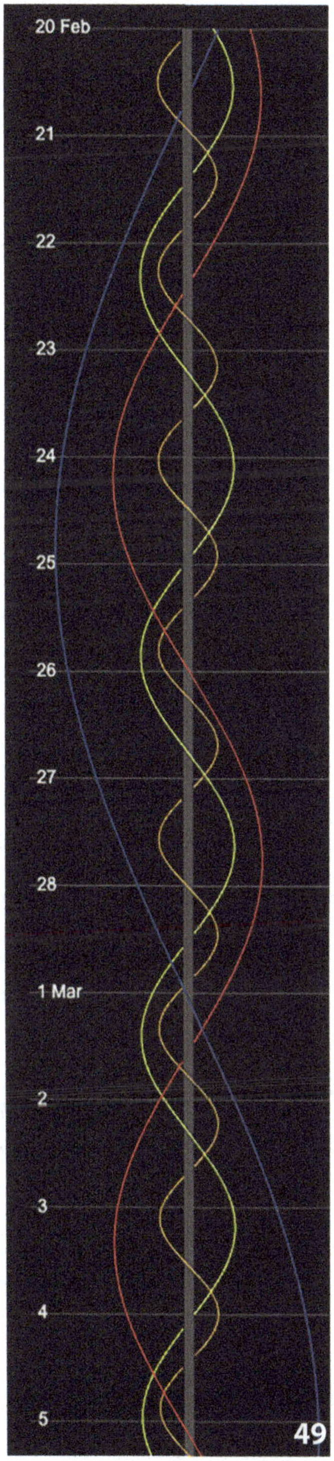

Helical diagrams like this show the positions of the Jupiter's moons from night to night and even hour to hour. These diagrams are found in astronomy magazines and, in this case, the website in-the-sky.org.

Saturn

Saturn is the most stunning object observable through a telescope. Being with someone when they see Saturn through a telescope for the first time is an incredible experience, and it's why so many astronomers do public outreach with their telescopes in the first place.

Saturn's glorious rings set this planet apart from the others. Through a small telescope on a clear evening, you may be able to see the gap between the two main rings, known as the Cassini Division.

On most nights, and with most telescopes, you should be able to spot Saturn's largest moon, Titan. Titan orbits about a million kilometres from Saturn (this is nearby in an astronomical sense). It is about half again as large as our Moon, and is covered in a thick methane atmosphere, giving it a yellowish–orange tint.

Depending on your telescope's aperture and the clarity of your skies, several more of its moons may also be visible. The three next-largest moons, Rhea, Dione, and Tethys, should be visible at apertures of about 5 inches and higher. At apertures of 10–12 inches, it becomes possible to observe the moons Iapetus, Hyperion, and Mimas.

When viewing Saturn at the eyepiece, it may be tempting to pull out your phone and try to capture an image. I've found this to be extremely challenging (and frustrating) without a tracking telescope and a premium phone adapter.

The most detailed images of Saturn came from NASA's Cassini spacecraft. One of the Cassini probe's most exciting finds occurred while investigating Saturn's moon Enceladus, where it photographed over one hundred erupting geysers. These geysers were shown to be spewing water and molecular hydrogen from a subsurface ocean into space. We'll have to wait until a future mission to investigate this occurrence in more detail, but this discovery adds yet another world where scientists may search for life.

Observing Log

Date: | **Time:** | **Location:**

Sky Conditions (Seeing/Transparency):

Instrument (Magnification/Aperture):

Notes:

Sketch

Saturn and its moons as they would appear through a small telescope

Object of Interest: Saturn
Dark Sky Requirement: City-viewable
Object Type: Planet (gas giant)
Brightness: −0.55 (brightest), +1.17 (dimmest)
Distance from the Sun: 1,433.53 million km
Apparent Size: 15 to 21 arcseconds (excluding rings)

Saturn's rings imaged by the Cassini spacecraft

On January 14, 2005, a probe called "Huygens," part of the Cassini mission, landed on Saturn's Moon Titan and captured this image of its surface.

Venus and its Phases

This beautiful planet is named after the Roman goddess of love and beauty. Venus is closer to the Sun than Earth. Therefore, you can see Venus only shortly after sunset or right before sunrise.

Venus is bright. So bright, in fact, that it is one of the primary sources of UFO sightings amongst pilots. This is due to an optical illusion. Large objects viewed at great distances appear to move along with an observer, creating the illusion that the observer is being followed by the object—in this case, Venus.

Through a telescope, Venus looks a bit like our Moon, although it is completely devoid of surface features. It appears white and even has phases. This is because Venus is closer to the Sun than Earth, and its nighttime side sometimes faces us.

Observing Log

Date:	Time:	Location:
Sky Conditions (Seeing/Transparency):		
Instrument (Magnification/Aperture):		
Notes:		

Sketch

Object of Interest: Venus
Dark Sky Requirement: City-viewable
Object Type: Planet
Brightness: −3.0 (dimmest) to −2.9 (brightest)
Distance: ~108 million km from the Sun
Apparent Size: 910 to 66 arcseconds (a massive swing)

Venus as it would appear through a small telescope

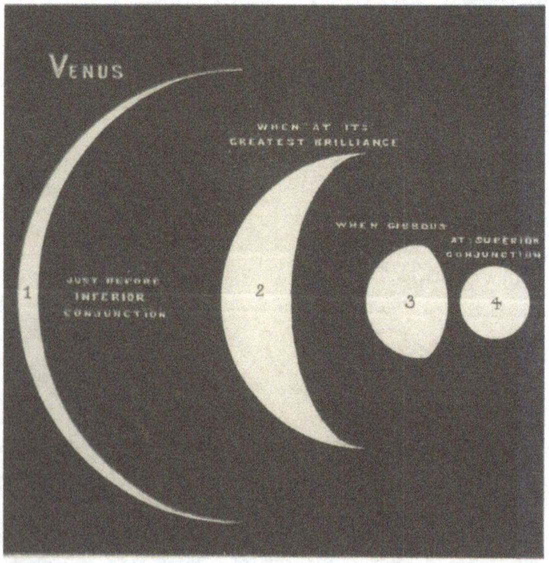

Description and drawing of the phases of Venus from An Atlas of Astronomy by Sir Robert Ball (1889)

Mars

It might look like just a small red disk in your telescope, but hey, it's Mars! With frequent observation and steady skies, you may be able to see the polar ice caps and some varying colors in the Martian soil.

This may be a good time to practice with higher-magnification eyepieces and to add the Barlow lens that may have come with your telescope. Adding a Moon filter or other filters that screw into the bottom of the eyepiece might also help reduce the planet's brightness or change the contrast enough to see detail on the surface.

In observing Mars, it's interesting to acknowledge that there are men and women here on Earth (at NASA's Jet Propulsion Laboratory in Los Angeles County) remotely piloting rovers the size of small SUVs on its surface. There are currently two operational rovers on Mars: Curiosity, which landed in 2012, and Perseverance, which landed in 2021.

Object of Interest: Mars
Dark Sky Requirement: City-viewable
Object Type: Planet
Brightness: –2.9 (brightest), 1.9 (dimmest)
Distance: ~250 million km from the Sun
Apparent Size: 3 to 25 arcseconds (a massive swing)

Mars as it would appear at high magnification at opposition under ideal seeing conditions

Observing Log

Date:	Time:	Location:
Sky Conditions (Seeing/Transparency):		
Instrument (Magnification/Aperture):		
Notes:		

Sketch

God of War

Mars gets its name from the Roman god of war, a god that shares his roots with the Greek god of war, Ares. Mars wasn't only regarded as a military god in Roman and Greek Cultures, but was also known as the war god Angaraka in East Asia. West Asian countries often referred to this planet as the "Fire Star," which also occasionally represented war.

The "Other" Planets

Mercury, Venus, Mars, Jupiter, and Saturn are visible without a telescope and have been observed by humans since ancient times. Uranus was the first planet ever "discovered." It was discovered by British astronomer William Herschel in 1781.

The orbit of Uranus was calculated soon after its discovery, but later, when its observed orbit didn't line up with calculations, it was predicted that another planet was influencing its path through space. This other planet, Neptune, was discovered about 60 years later.

Planets change their positions from night to night, so you'll need astronomy software to predict their location. Planets that are close to the Sun move faster than those farther away. That's why Mercury is only visible for a few days and Venus is visible for several weeks at a time. The remaining planets stay in the evening sky for about half the year.

Occasionally, the outer planets (Mars in particular) will appear to move backward relative to the background stars. This occurs when Earth, which is moving faster, passes the planet in its orbit, just like how cars appear to go backward (relative to your car) when you overtake them on the highway.

Mercury is bright but still challenging to observe. It is only possible to view this planet for a week or so, several times each year. And on these nights, it's only visible for a few minutes after sunset or before sunrise.

Uranus and Neptune will appear as small bluish points, almost like stars. Larger telescopes can resolve these into small blue-green disks. Pluto is nearly impossible to identify in a small telescope. Even in a large telescope, it will only appear as a very dim star.

I've added the minor planet (asteroid) Vesta because it is typically slightly brighter than Neptune. Vesta will appear as a white star of moderate brightness. If you take a photo of it one night and then again on the next (I'd recommend using a smart telescope), you will see that it will have moved against the background stars.

Observing Log

Date:	Time:	Location:

Sky Conditions (Seeing/Transparency):

Instrument (Magnification/Aperture):

Notes:

Sketch

Object of Interest: Mercury
Brightness: –2.48 (brightest) to +7.25 (dimmest)
Distance from the Sun: 57.91 million km
Apparent Size: 4.5–13.0 arcseconds

Mercury imaged in representative color by the MESSENGER spacecraft

Object of Interest: Uranus
Brightness: about 5.7
Distance from the Sun: 2.9 billion km
Apparent Size: 3.3 to 4.0 arcseconds

Uranus imaged by the Voyager 2 spacecraft

Object of Interest: Neptune
Brightness: about 7.8
Distance from the Sun: 4.50 billion km
Apparent Size: about 2.3 arcseconds

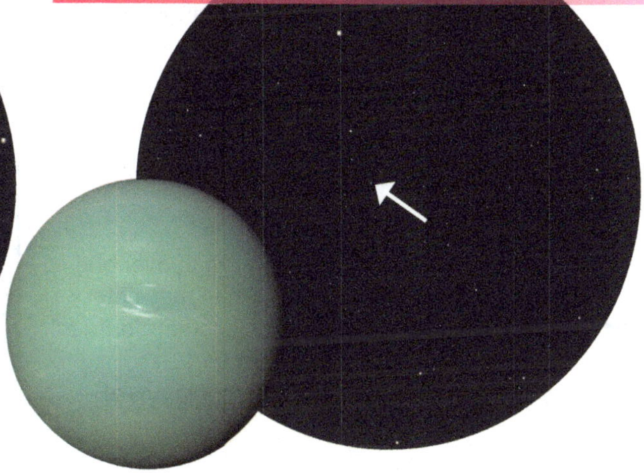

Neptune imaged by the Voyager 2 spacecraft

Object of Interest: Asteroid Vesta
Brightness: 5.1 (brightest) to 8.5 (dimmest)
Distance from the Sun: 353 million km
Apparent Size: ~0.5 arcseconds (starlike)

Asteroid Vesta imaged by the Dawn spacecraft. This asteroid is visible in a telescope as a small point of light.

Object of Interest: Pluto
Brightness: 14.5 (VERY dim)
Distance from the Sun: ~5 billion km
Apparent Size: 0.1 arcseconds (starlike)

Pluto imaged by the New Horizons spacecraft and by the Burke-Gaffney Observatory

Comets

Comets visible in the night sky usually get picked up by the media. However, articles often make vastly exaggerated claims of brightness. Despite the hype, only a few of these comets can actually be glimpsed by the casual stargazer. Comets are the only celestial objects that can be named after their discoverers, although these days most are discovered by robotic telescopes and take on the name of the instrument itself.

Comets are city-sized balls of ice, often traveling over one hundred thousand kilometres per hour. When they enter the inner Solar System where Earth resides, comets "outgas" as they come closer to the Sun and release particles and gases, creating a visible tail that is often millions of kilometres long.

We observe comets from a distance of hundreds of millions of kilometres. Although they are traveling at great speed, they are often visible for about a month. This gives the amateur astronomer plenty of time to observe them.

Astronomy websites will usually report when a comet is visible in the night sky. However, I've found the maps they provide are often too difficult to read. Bringing the location of the comet up in stargazing software may make them much easier to find.

You may want to use binoculars to make the first observation, then switch to your telescope for a closer look. It is possible to photograph bright comets in the sky with a good camera on a tripod, as in the photo at right.

Comet 67P imaged by the Rosetta spacecraft

Observing Log

Date:	Time:	Location:
Sky Conditions (Seeing/Transparency):		
Instrument (Magnification/Aperture):		
Notes:		

Sketch

Comet C2023 A3
Image taken from the city with a Seestar S50 (right, with the comet in view)

Adding Recent Comets to Stellarium (on Mac/PC)

For a newly discovered comet, the orbit may need to be downloaded, and the software updated.
To do this, select:

- Configuration>Plugins>Solar System Editor
- Configure> "Solar System"> Import Orbital Elements in MPC Format
- Select "Comets," select "Gideon van Buitenen: comets," and then click "Get Orbital Elements"
- Search for the comet by name, select it, and click "Add Objects."

Comet finding charts provided online usually aren't very intuitive. I've found it easier to make my comet map with Stellarium.

Once you know where the comet is in the sky (or at least its approximate location), you might want to first search for it using binoculars. Finding it with a telescope may be too challenging due to the limited field of view. After you've found it in binoculars, only then would you move over to the telescope.

Solar Observing

The Sun can be observed through a telescope with a proper solar filter covering the aperture. The Sun is an extremely dynamic system with constantly changing features. In a beginner's telescope, you're generally looking at sunspots. To see surface detail such as prominences (resembling giant flames), you need a specialized solar telescope such as a Coronado or Lunt Solar Telescope. These specialized solar telescopes only observe certain wavelengths of light, and are often called hydrogen-alpha (or H-alpha) telescopes.

Sunspots are eddies or storms of magnetic activity near the surface of the Sun. These storms reduce the surface temperature over a given area, and create dark spots when viewed through solar-filtered telescopes.

Sunspots are usually larger than the diameter of Earth and often come in pairs (one for each magnetic pole of the disturbance). Third, they change location every day, which is mainly due to the rotation of the Sun.

With the Sun in focus, you should almost always be able to see at least one or two sunspots. However, sunspot activity follows an 11-year cycle. During what we call "solar minimum," you may not see any sunspots at all for weeks or months.

Remember, never look at the Sun without an approved solar filter over the primary mirror or lens (not the eyepiece) of your telescope. Eyepiece solar filters exist, but I do NOT recommend you try these, as they are extremely risky.

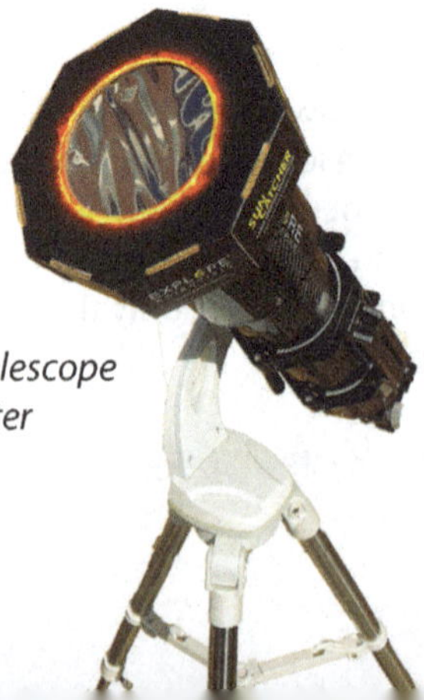

A 102-mm telescope with solar filter

Observing Log

Date:	Time:	Location:
Sky Conditions (Seeing/Transparency):		
Instrument (Magnification/Aperture):		
Notes:		

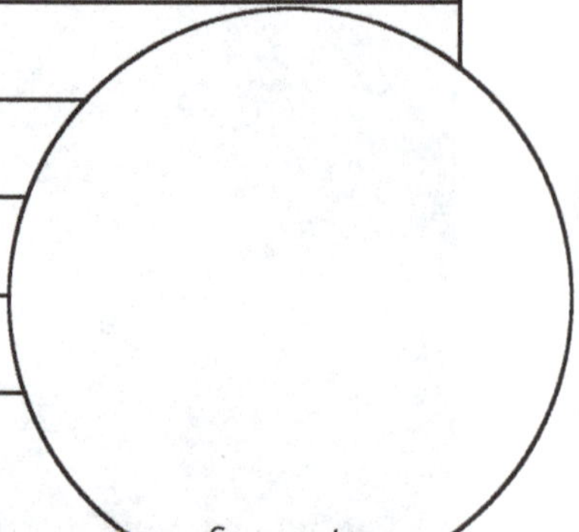

Sunspots

Warning! Pointing unfiltered optics at the Sun will cause permanent eye damage

Object of Interest: The Sun
Gear requirement: Solar filters
Object Type: G-type star
Brightness: –26.7
Distance: ~ 149 million km
Apparent Size: 30 arcminutes

Stella helping me take this photo

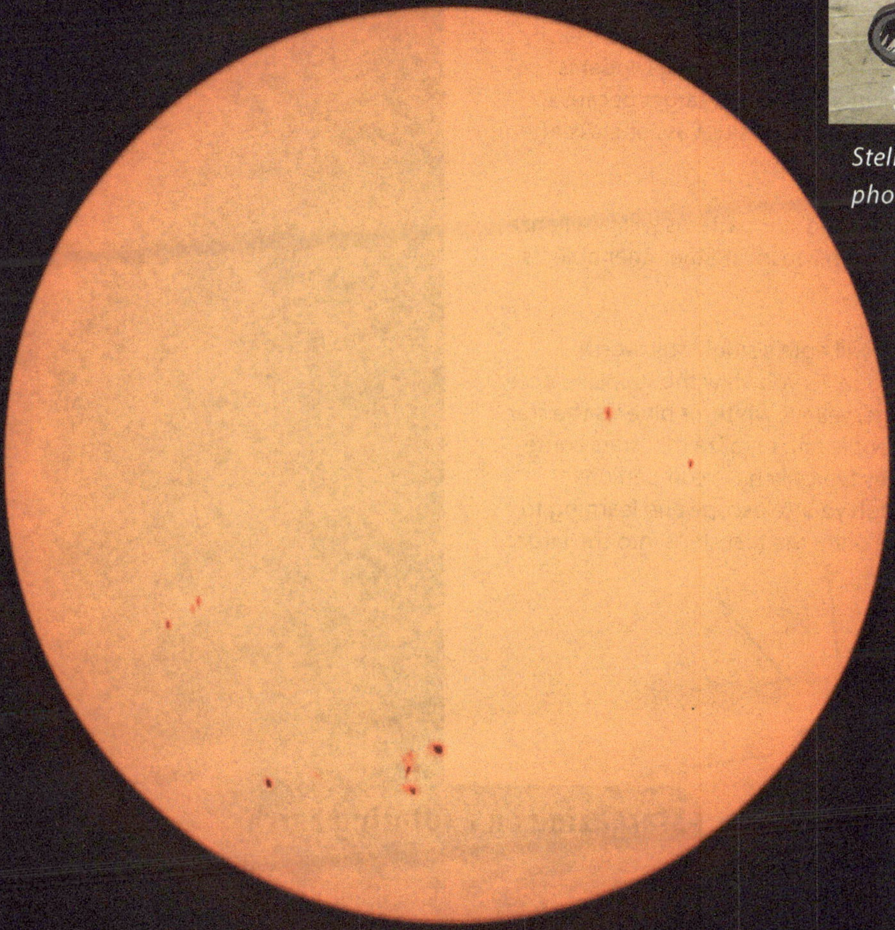

Most Smart Telescopes come with a solar filter. Try it out!

Circumpolar Targets

Objects That Never Set From Mid-Northern Latitudes

This section will cover the most popular circumpolar targets visible year-round from mid-latitudes in the Northern Hemisphere. That said, there are times of the year when these objects are higher in the sky and therefore more easily visible than at other times. For example, the Big Dipper is often considered a spring or early summer target because this is when it appears overhead in the sky. Cassiopeia is often considered an autumn or winter target.

Learning to locate these stars and star patterns will familiarize you with the night sky and help you locate the other objects mentioned in this book.

Some sections of this book will note a single star worth observing with your telescope. As you view the star, take note of its color—is it red, orange, yellow, white, or blue? Is the star alone or in a pair? Many people don't realize that stars come in several colors or that they typically have companions. Simply viewing stars through your telescope and learning to appreciate their beauty are often the first steps into the larger universe.

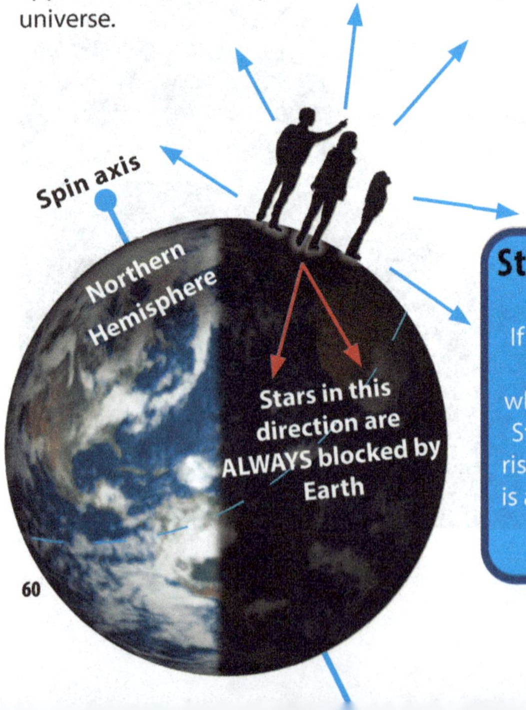

Stargazing on a rotating Earth

If you live in the Northern Hemisphere, there is a region of sky to the north where the stars rotate around the North Star and never set. The remaining stars rise in the east and set in the west. There is also a region of sky to the south that is always blocked by Earth.

A ten-minute exposure while looking north. This effect is called "star trails" and is caused by sky rotation.

Circumpolar Sky at a Glance

The north circumpolar sky is dominated by two easily recognizable star patterns: the Big Dipper, an asterism within the constellation Ursa Major, and the Big W, five bright stars within the constellation Cassiopeia.

About once per day, and once per year, the stars appear to rotate around the North Star (Polaris), which appears fixed in the sky. It's just a coincidence that there is a relatively bright star located near the North Celestial Pole. The South Celestial Pole contains no such bright star.

It helps to be able to recognize some of the lesser known constellations here, since these will help guide you to the targets within. Camelopardalis, for example, contains, Kemble's Cascade (page 70).

Hazy view of the Big Dipper from Stargaze Nova Scotia

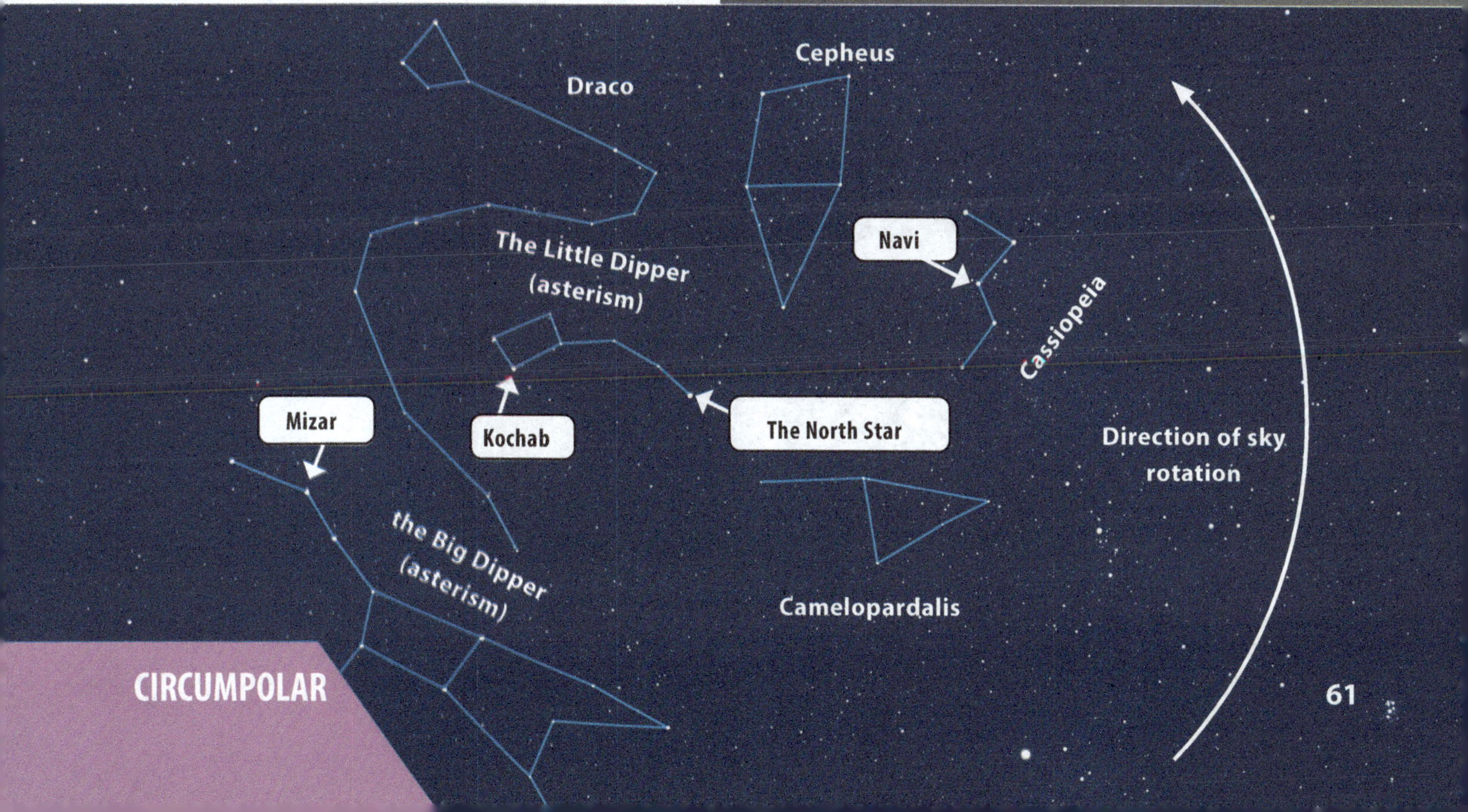

North Star (Polaris)

There's a popular misconception that the North Star is the brightest star in the sky. In reality, the North Star is around the forty-eighth brightest star in the night sky.

There are two things to look for when viewing Polaris. First, it is a double star whose dim companion can be seen in a moderately sized telescope. Second, Polaris forms the diamond in the Engagement Ring asterism, a tight circle of 7 stars, best viewed at low magnification.

To find the North Star, follow the two stars (often called the Pointer Stars) that form the front of the cup of the Big Dipper to the next brightest star (as shown in the diagram below).

Polaris is very important to observers who own an EQ mount in the Northern Hemisphere. In order for this type of mount to function, one axis must be pointed very near this star.

Observing Log

Date:	Time:	Location:
Sky Conditions (Seeing/Transparency):		
Instrument (Magnification/Aperture):		
Notes:		

Sketch

Object of Interest: The North Star
Dark Sky Requirement: City-viewable
Object Type: Double star
Brightness: 2.0
Distance: 430 light years
Apparent Size: Point source

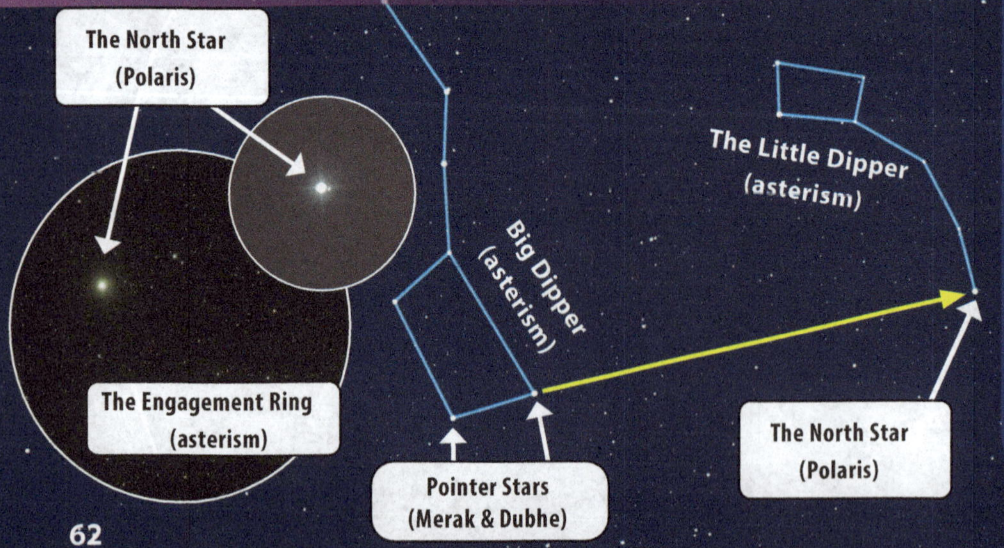

Wayfinding with the stars

Polynesian voyagers explored the Pacific Islands, navigating by the winds and stars thousands of years before Europeans sailed to North America. These voyagers had to memorize the locations of hundreds of stars along with knowledge of the winds and currents.

Voyagers use a compass with 32 directional points. When a navigation star rises or sets, its location is known to fall along one of these points, and they can orient their vessel accordingly.

Mizar & Alcor

There is no need for an optometrist when you have these two stars in sight. Nicknamed the "Horse and Rider," seeing these stars, located in the Big Dipper, used to be a test of eyesight!

Mizar and Alcor form the center of the handle of the Big Dipper. When observing these stars with a telescope, Mizar is a star with a companion so close they appear to almost touch, giving them the nickname "the Snowman".

In addition to being a visual double star, Mizar was the first telescopic double star ever observed. It was discovered by an Italian mathematician named Benedetto Castelli in 1617.

Double star observing is a hobby in itself for many amateur astronomers. Hundreds of popular double stars are visible in small telescopes, and astronomers enjoy the challenge of splitting these stars into pairs.

Observing Log

Date:	Time:	Location:
Sky Conditions (Seeing/Transparency):		
Instrument (Magnification/Aperture):		
Notes:		

Sketch

Object of Interest: Mizar
Dark Sky Requirement: City-viewable
Object Type: Double star
Brightness: 2.2
Distance: 85.8 light years
Apparent Size: Point source

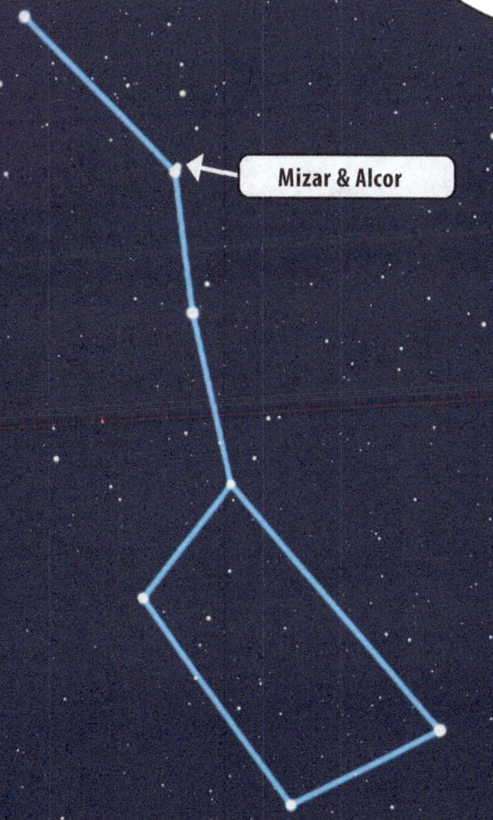

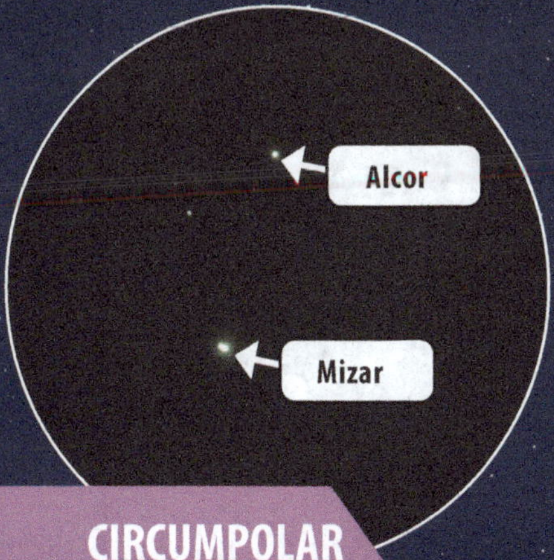

CIRCUMPOLAR

Double Cluster (NGC 869/884)

The double cluster is a great low-magnification target for telescopes and binoculars alike. The clusters are separated by about the 30 arcminutes, the apparent diameter of the Full Moon.

These two clusters, designated NGC 869 and NGC 884 (or Caldwell 14, together), are visible to the unaided eye in extremely dark skies (a rarity in today's modern world).

Through a telescope, at low magnification, I look for the "crab claw"-like structure first, then move over to the clusters themselves using **averted vision** to maximize the number of visible stars.

To find the double cluster in Perseus, look to Cassiopeia (the Big W) and find the clusters below and to the left of the W. I typically view this on late summer evenings when it's rising in the northeastern sky.

Observing Log

Date:	Time:	Location:
Sky Conditions (Seeing/Transparency):		
Instrument (Magnification/Aperture):		
Notes:		

Sketch

Object of Interest: NGC 869
Dark Sky Requirement: Suburbs
Object Type: Open cluster
Brightness: 3.8
Distance: 6,781 light years
Apparent Size: 30 arcminutes

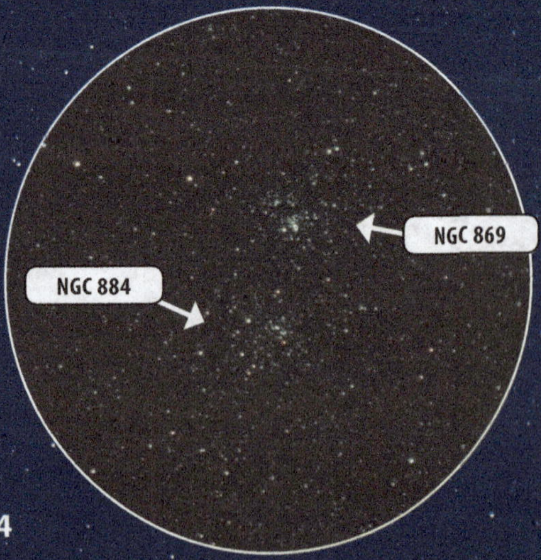

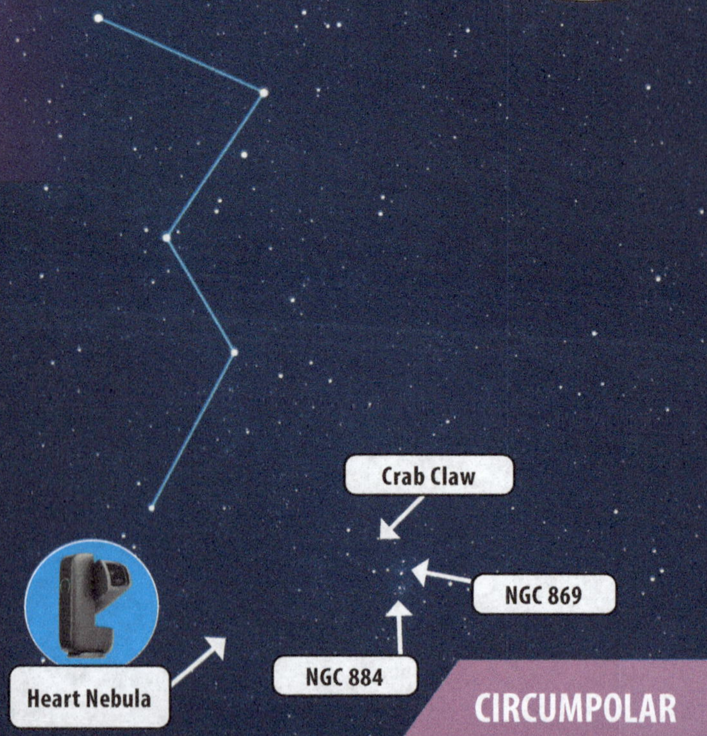

Dragonfly Cluster (NGC 457)

This fantastic open cluster in Cassiopeia used to be primarily known as the Dragonfly Cluster (or the Owl Cluster). After the Stephen Spielberg classic *E.T. the Extra-Terrestrial* hit theaters in the spring of 1982, however, this cluster underwent a colloquial name change to "The E.T. Cluster" … and it's easy to see why.

You have to see this one to believe it. Photographs like the one on the right simply don't capture the effect of two bright eyes and arms reaching out to grab you!

I have viewed this cluster regularly with students and at star parties in telescopes of all sizes, and it has since become one of my favorite objects to view in the night sky.

The E.T. Cluster is easy to find due to its proximity to the Big W (Cassiopeia).

Observing Log

Date:	Time:	Location:
Sky Conditions (Seeing/Transparency):		
Instrument (Magnification/Aperture):		
Notes:		

Sketch

Object of Interest: NGC 457
Dark Sky Requirement: Suburbs
Object Type: Open star cluster
Brightness: 6.4
Distance: 7,900 light years
Apparent Size: 20 arcminutes

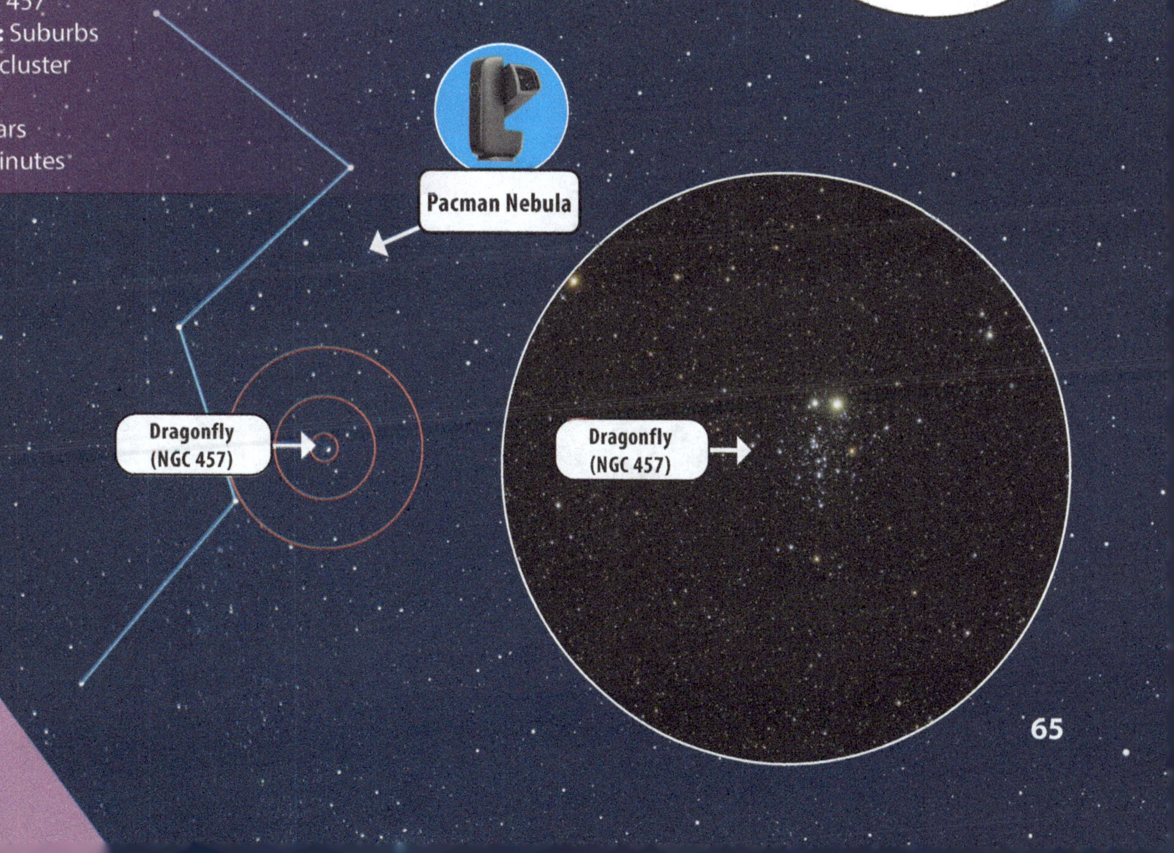

CIRCUMPOLAR

Whirlpool & Pinwheel Galaxies (M51 & M101)

The Whirlpool Galaxy, or M51, is easy to find in a small telescope, or even binoculars, but only on moonless nights, I've even been able to observe it from the city. This galaxy is accompanied by a smaller companion galaxy designated NGC 5195, or M51b. The gravitational interaction between these two structures is thought to give the Whirlpool its well-defined whirlpool shape.

I've had less luck observing the Pinwheel Galaxy, or M101, from the city. From dark skies, however, like those in Yosemite National Park, I was easily able to see a fair amount of detail within the spiral arms themselves!

There are several other galaxies in this part of the sky, near the handle of the Big Dipper. However, I've found that M51 and M101 are by far the easiest to find in a small telescope. The best way to observe these other galaxies, M62, M94, and M106, is to find yourself in extremely dark skies (i.e., Bortle 1 or 2).

Observing Log

Date: **Time:** **Location:**

Sky Conditions (Seeing/Transparency):

Instrument (Magnification/Aperture):

Notes:

Sketch

Object of Interest: The Pinwheel Galaxy (M101)
Dark Sky Requirement: Dark, moonless skies
Object Type: Spiral galaxy
Brightness: 7.8
Distance: 23 million light years
Apparent Size: 24 × 23 arcminutes

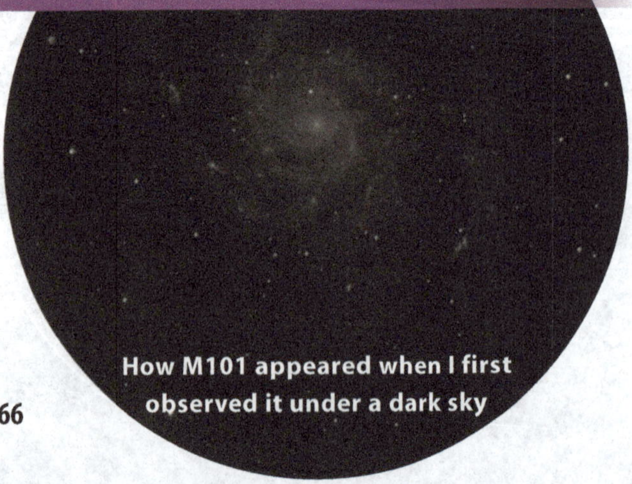

How M101 appeared when I first observed it under a dark sky

Object of Interest: The Whirlpool Galaxy (M51)
Dark Sky Requirement: Dark, moonless skies
Object Type: Spiral galaxy
Brightness: 8.4
Distance: 31 million light years
Apparent Size: 11 × 7 arcminutes

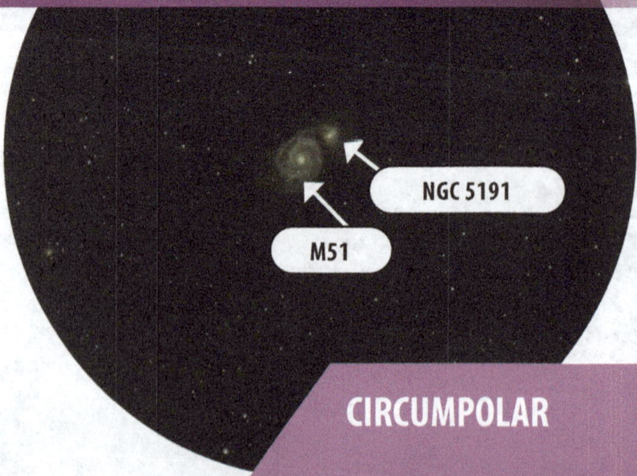

CIRCUMPOLAR

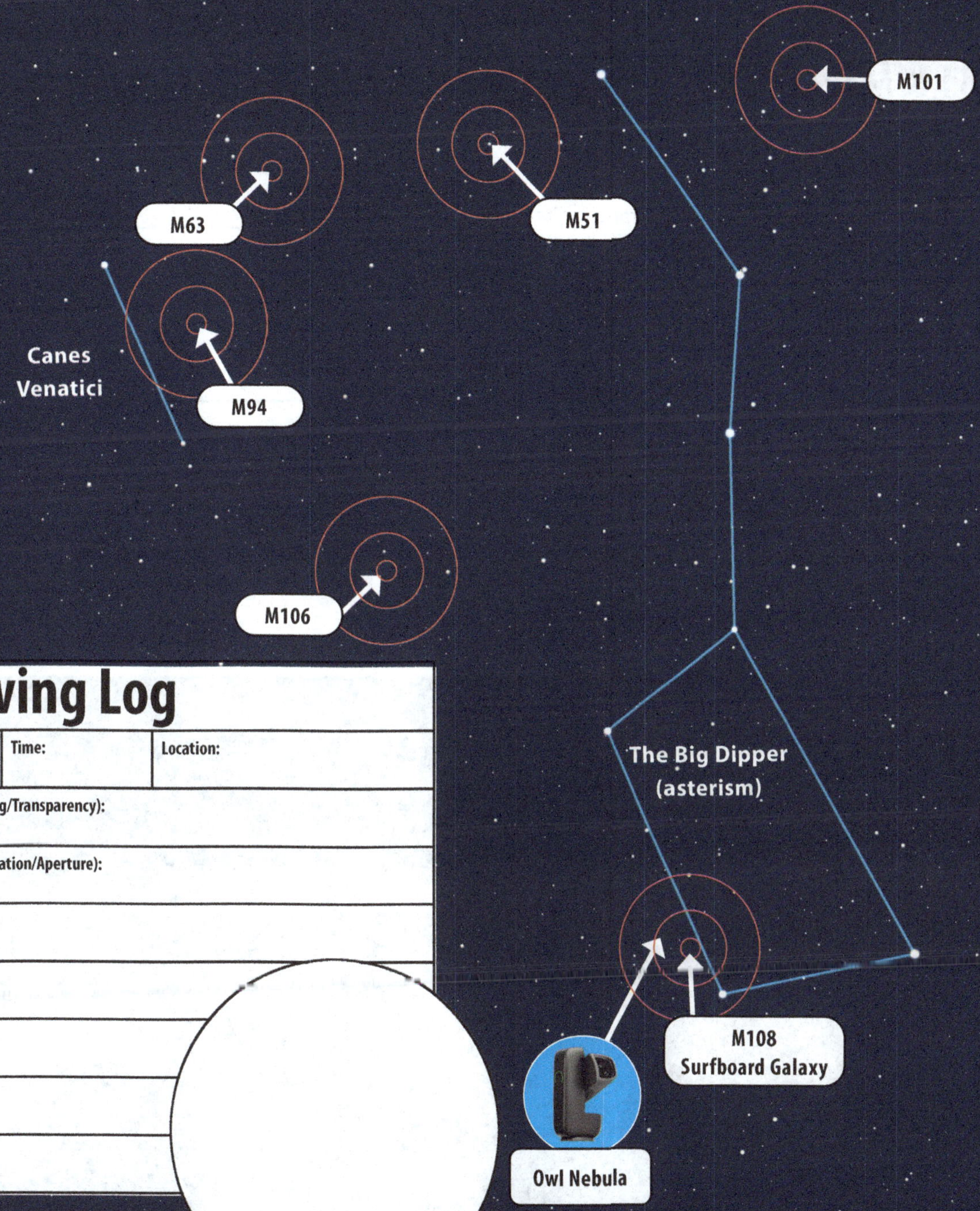

Observing Log

Date:	Time:	Location:

Sky Conditions (Seeing/Transparency):

Instrument (Magnification/Aperture):

Notes:

Sketch

Bode's & Cigar Galaxies (M81 & M82)

After Andromeda, I have found that M81 and M82 are the two galaxies that are the easiest to find.

M81, referred to as Bode's Galaxy, was named for its discoverer: the German astronomer Johann Elert Bode. Bode can also be blamed for suggesting the name for the planet Uranus, edging out other names like Cybele, Herschel's Planet, and Georgium Sidus. M82 is commonly referred to as the Cigar Galaxy because of how it appears from Earth (it is a spiral galaxy, viewed edge-on). To view these galaxies, use an eyepiece with low magnification. With the Big Dipper as a guide, draw a line between the bottom left of the Dipper's bowl and its lip. Then, extend this line from the lip to arrive at the location of these galaxies.

In images that include M82, like those on the right taken with a small telescope, you can see jets of gas emanating from its galactic core. This gas seems to be primarily composed of hydrogen, which gives it a pink color in photos.

Observing Log

Date:	Time:	Location:
Sky Conditions (Seeing/Transparency):		
Instrument (Magnification/Aperture):		
Notes:		

Sketch

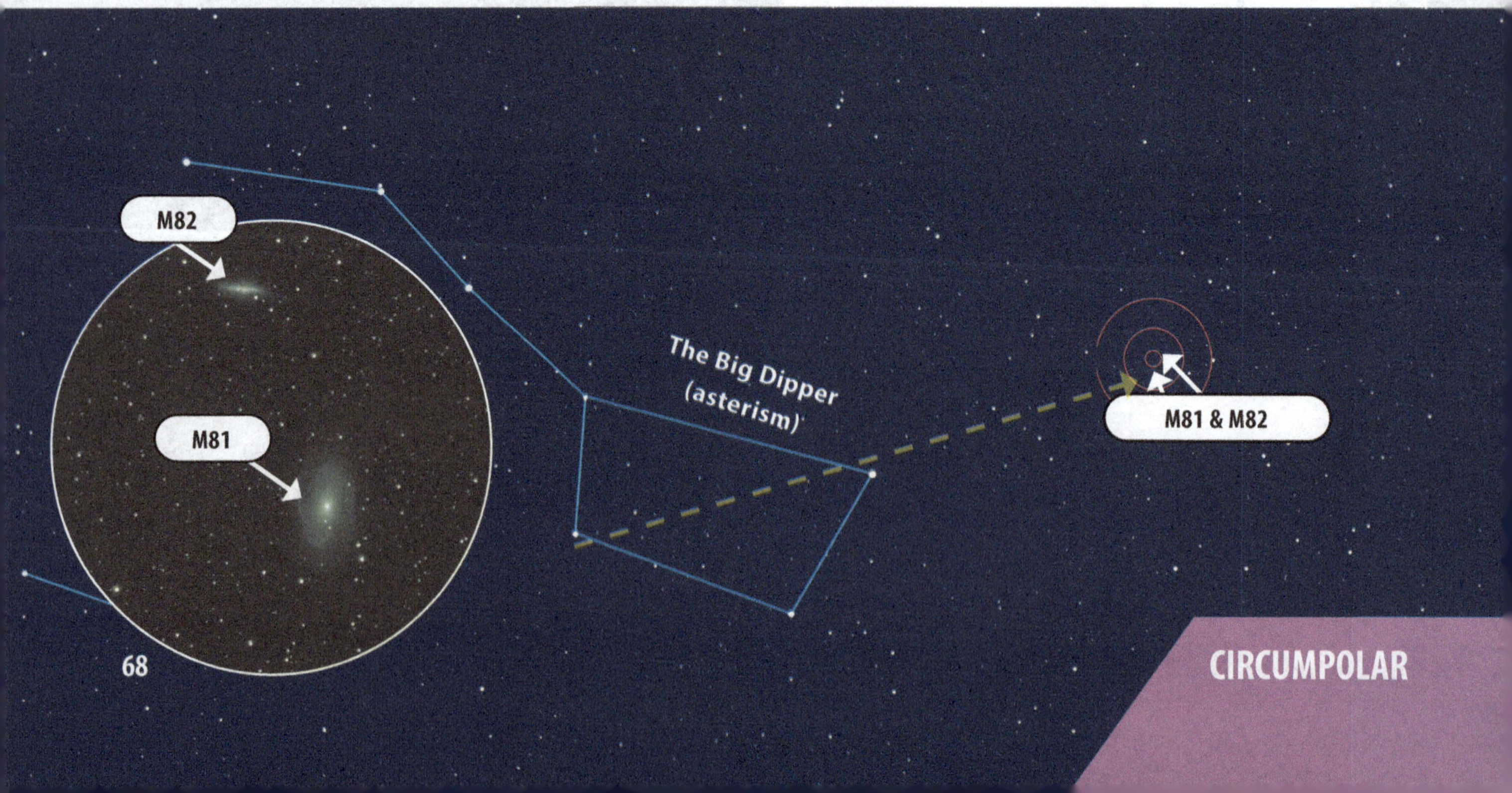

Object of Interest: The Cigar Galaxy (M82)
Dark Sky Requirement: Suburbs
Object Type: Starburst/spiral galaxy
Brightness: 8.0
Distance: 12 million light years
Apparent Size: 11 × 5 arcminutes

Object of Interest: Bode's Galaxy (M81)
Dark Sky Requirement: Suburbs
Object Type: Spiral galaxy
Brightness: 6.8
Distance: 12 million light years
Apparent Size: 21 × 11 arcminutes

Kemble's Cascade & Jolly Roger Cluster (NGC 1502)

Kemble's Cascade is an extremely popular target in the amateur astronomy community. This line of colorful stars stretches across multiple telescope fields of view. This star group was popularized in 1980 by Father Lucien J. Kemble, a Canadian Franciscan priest and avid amateur astronomer

The cascade terminates in an open star cluster called NGC 1502, nicknamed the Jolly Roger Cluster. It gets this nickname for its resemblance to the pirate flag, with a star grouping in the center representing the skull and the stars in the four corners representing the crossbones.

The host constellation, Camelopardalis, can be a bit tricky to find, as its stars are quite dim. I find the cascade in relation to Cassiopeia, which, in the image below would appear as an "M" on an autumn or winter evening.

Observing Log

Date:	Time:	Location:

Sky Conditions (Seeing/Transparency):

Instrument (Magnification/Aperture):

Notes:

Sketch

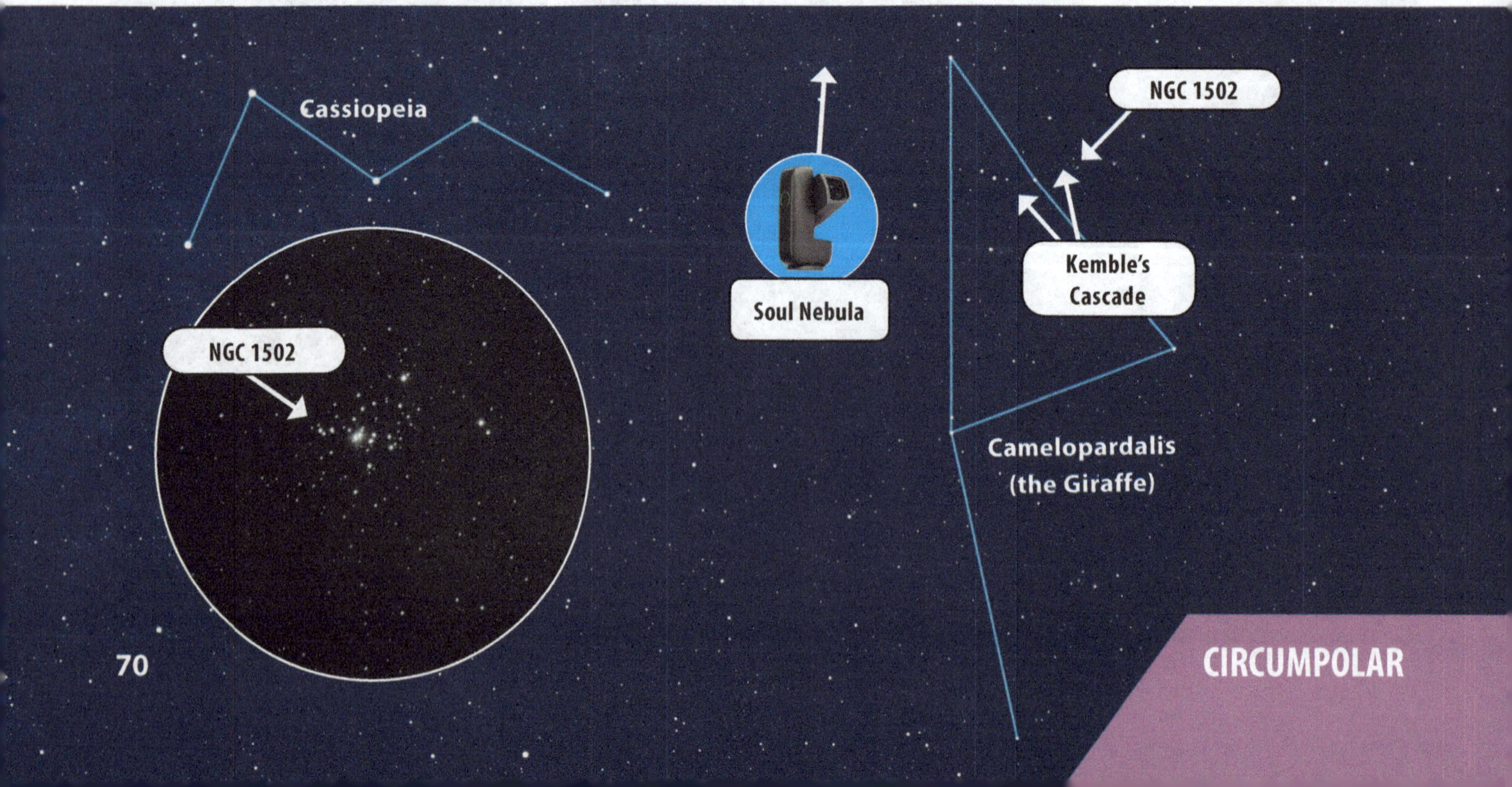

Object of Interest: NGC 1502
Dark Sky Requirement: Suburbs
Object Type: Open cluster
Brightness: 6.9
Distance: 3,300 light years
Apparent Size: 8 arcminutes

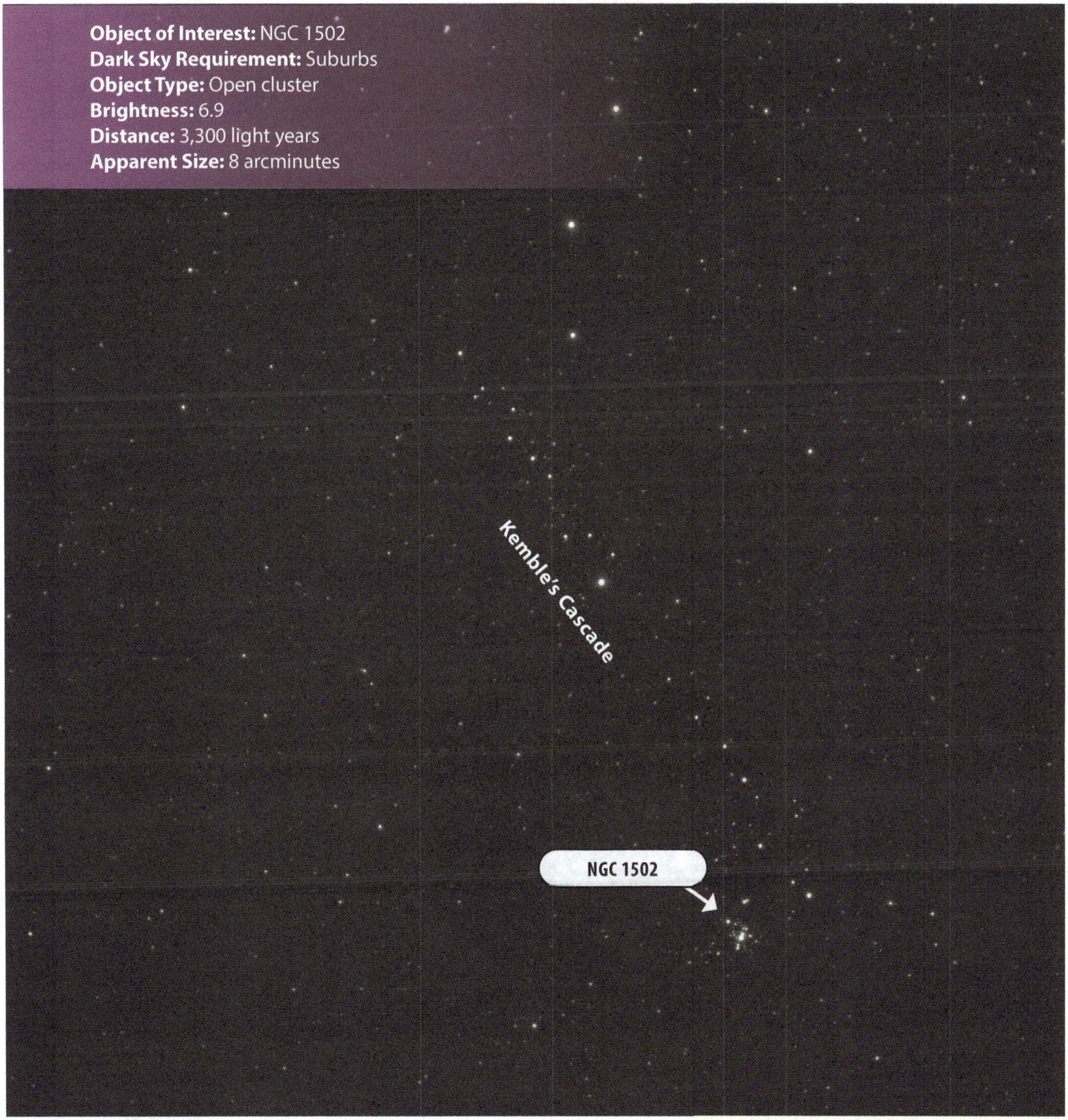

Circumpolar Double Stars and Asterisms

We mentioned double stars like Mizar and Alcor, as well as Polaris, in previous sections, but there are several other double stars of note in the circumpolar sky. Astronomy clubs often have special observing lists just for observing double stars.

There are two sets of "doubles" viewable with binoculars or small telescopes in Draco, listed in RASC's Explore the Universe observing program (https://www.rasc.ca/explore-universe). This program provides new observers with a certificate for observing the most popular targets in the sky. These stars are known as 16 and 17 Draconis, Nu 1 and Nu 2 Draconis, and Delta Cephei.

Some of these double stars are quite close together. Amateur astronomers speak of "splitting" doubles, and doing so successfully often relies on several factors. The seeing conditions, particularly turbulence in the air, can vary considerably, and on nights where seeing is poor, double stars will be much harder to split.

Asterisms are "unofficial" patterns of stars … and there are many to find! The RASC has a project called the "World Asterism Project" led by amateur astronomer and past president of the RASC, Charles Innis. Its first volume is currently 1,701 pages, and covers hundreds of different sky cultures.

One interesting grouping of stars has been nicknamed "the Shark," which can be found next to the cup in the Little Dipper. Another pattern of stars between Cassiopeia has been nicknamed "the Airplane." Open cluster 7160 was a favorite when I was doing public tours at the Burke-Gaffney Observatory, and it's been nicknamed "the Swimming Alligator." Can you find these asterisms?

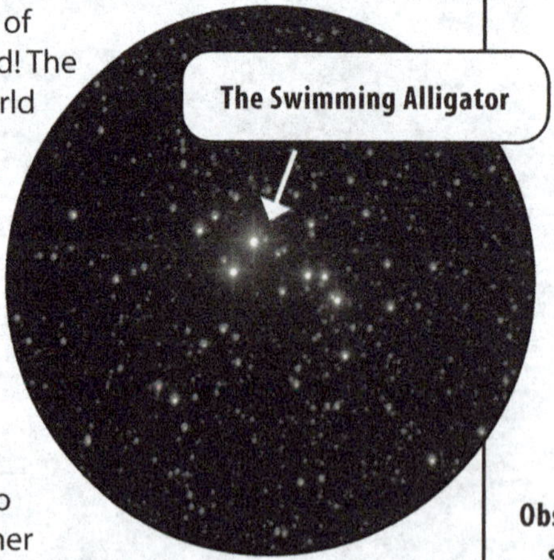

Observing Log

Date:	Time:	Location:

Sky Conditions (Seeing/Transparency):

Instrument (Magnification/Aperture):

Notes:

Sketch

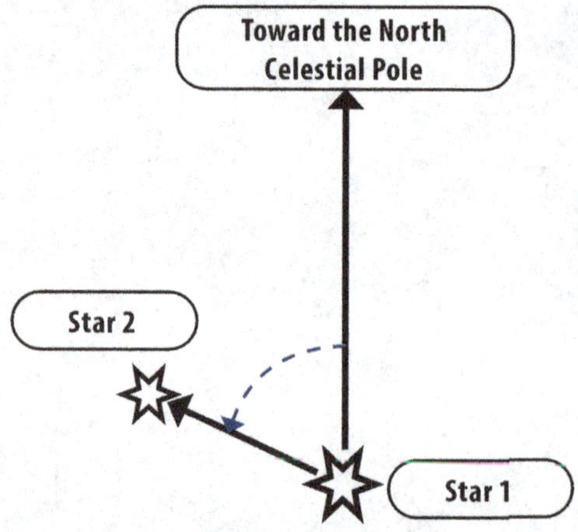

Observing programs for double stars typically list the separation (in arcseconds) and the "position angle" (PA), the counterclockwise angle between north (toward the North Star) and the companion (less bright) star.

The numbers below each double star represent the separation between the stars (symbols: ′ means arcminutes; ″ means arcseconds).

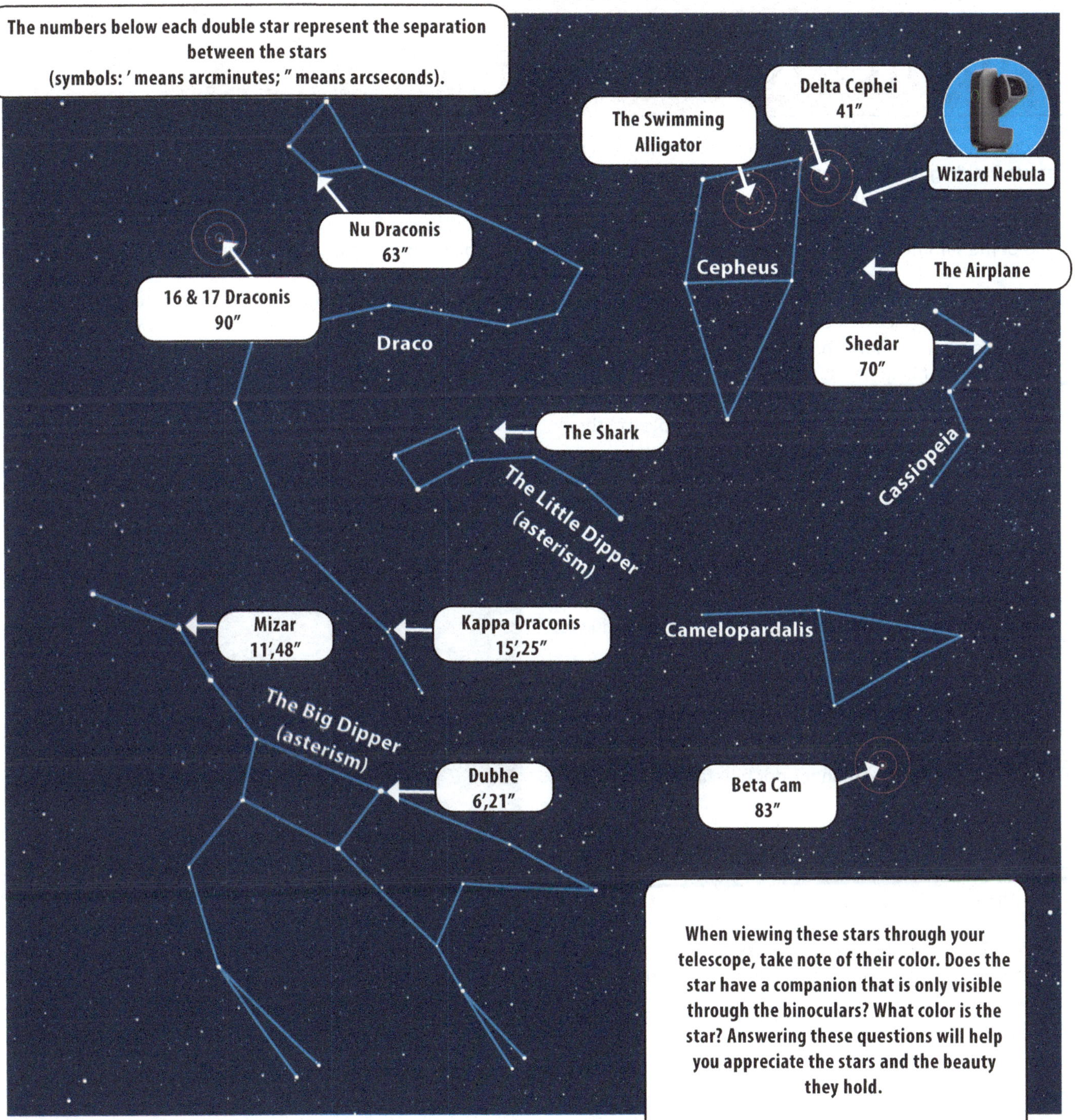

When viewing these stars through your telescope, take note of their color. Does the star have a companion that is only visible through the binoculars? What color is the star? Answering these questions will help you appreciate the stars and the beauty they hold.

Summer Targets

If you are far from the Equator, it really doesn't get dark until around midnight early in the summer. The map on the right shows how the sky would appear at around midnight at the beginning of summer, but at a couple hours earlier near the end of the summer.

August is when summer stargazing is at its peak. It gets dark earlier, and once it does, the Milky Way rises straight up out of the southern horizon. Sagittarius, where you'll find the core of the Milky Way, is perfectly positioned to allow observing of some of the summer's most exciting deep-sky targets. If a New Moon coincides with the peak of the Perseid meteor shower, you're really in for a show.

The summer sky can be divided into two separate regions. As a stargazer, this is where you'll spend most of your time. First, there's the area surrounding the Teapot asterism in Sagittarius, a region of sky which rises and sets low on the southern horizon. Second, there's the Summer Triangle, which sticks around almost until Christmas due to lengthening nights.

The area near the Teapot contains some of the brightest nebulae, such as the Lagoon, the Swan, and the Trifid, while the Summer Triangle contains a variety of different targets such as the Coathanger, double star Albireo, and the Dumbbell Nebula.

The Summer Triangle is a great way to learn your way around the sky. It is outlined by three stars: Vega, Deneb, and Altair. Use the acronym DAVe to remember them.

An asterism called the Northern Cross (in the constellation Cygnus) is often referenced to locate objects in this part of the sky.

Vega is the brightest star in the Summer Triangle. Interestingly, Vega served as the North Star about fourteen thousand years ago, and it will be again about twelve thousand years from now. This is due to a slow wobbling of the polar axis of Earth, called precession.

Altair is frequently mentioned in science fiction due to its proximity to Earth. At only 16.7 light years away, it is one of the closest bright stars. Altarian dollars is the currency used throughout *The Hitchhiker's Guide to the Galaxy*. Altair is also mentioned in multiple *Star Trek* TV show episodes, *Star Trek II: The Wrath of Khan*, and also two episodes of *Doctor Who*.

In extremely dark skies, such as at national parks, you can scan the skies for "dark nebulae," defined as dark regions of space dust that obscure the background stars. These are best observed at low magnifications (or even without a telescope) near the core of the Milky Way.

Image: Dark nebula, Antares (star), and M4 (globular cluster)

The Summer Sky at a Glance

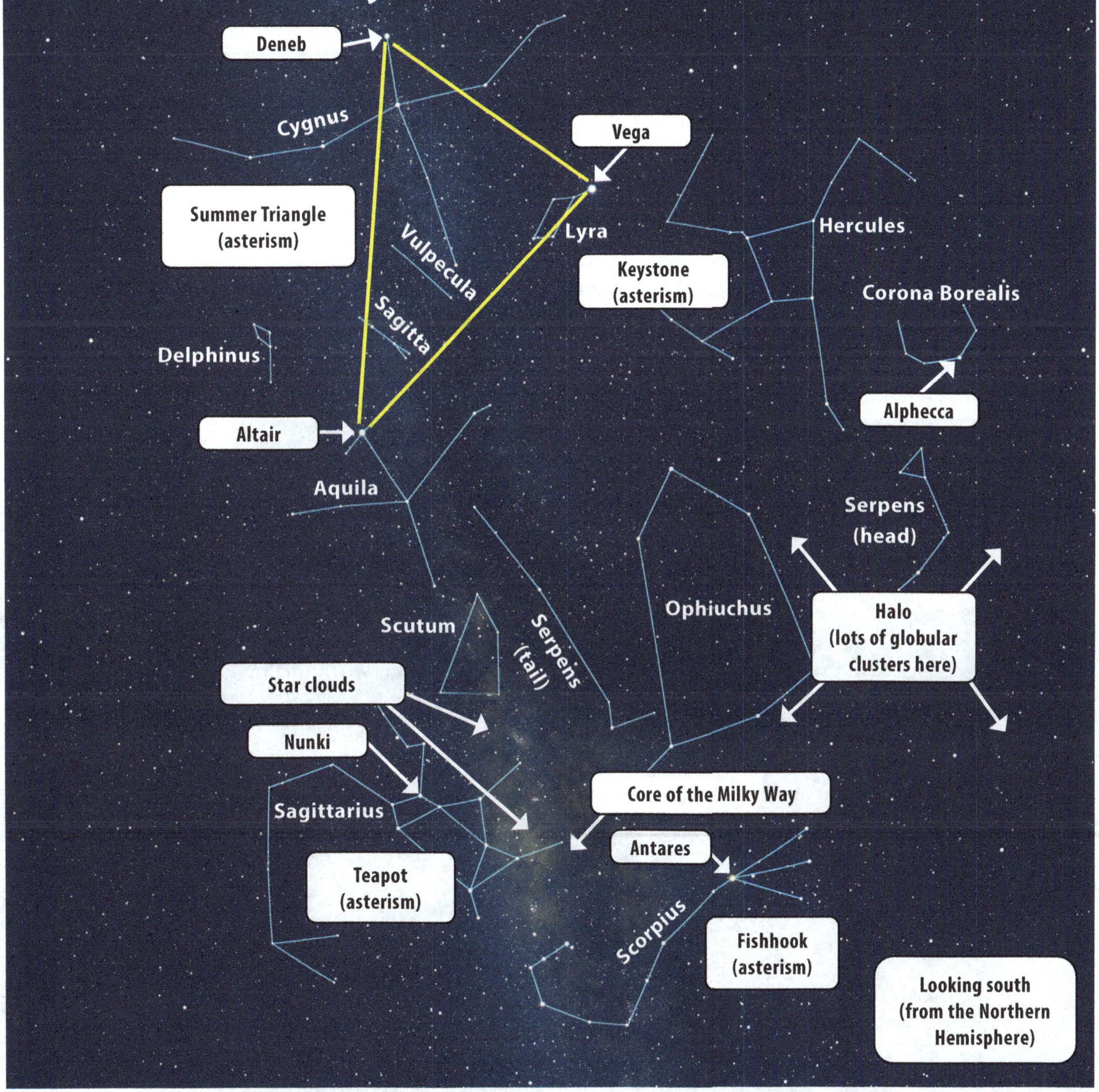

Summer Beehive (IC 4665)

The Summer Beehive is also known as the "Hi" cluster, as its star pattern appears to spell the word. Early in the summer, the "original" Beehive Cluster may still be visible in the center of the springtime constellation, Cancer.

Sky & Telescope contributor Massimo Difusco calls this "The Neglected Star Cluster" as it is one of the most prominent clusters excluded from the famous Messier and Herschel lists.

Ophiuchus lies in the direction of our galaxy's halo, a region of the sky filled with globular star clusters such as M10 and M12. Most of the globular star clusters are quite distant and small, but in the darkest skies, a keen observer should be able to pick out several of these clusters in a telescope.

Observing Log

Date:	Time:	Location:
Sky Conditions (Seeing/Transparency):		
Instrument (Magnification/Aperture):		
Notes:		

Sketch

Object of Interest: The Summer Beehive (IC 4665)
Dark Sky Requirement: City-viewable
Object Type: Open cluster
Brightness: 4
Distance: 1,100 light years
Apparent Size: 70 arcminutes

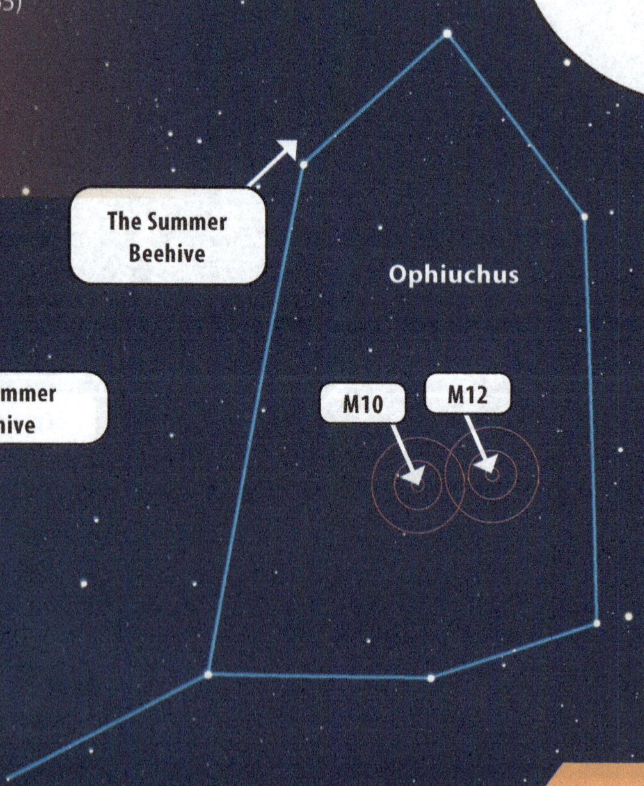

The Spider Cluster (M4)

The Spider Cluster, M4, is one of the easiest globular clusters to find due to its proximity to the bright red star Antares, which is part of an open cluster named Collinder 302. M4 appears within the cluster, despite being many times farther away.

Through a telescope, M4 doesn't look as impressive as M13 or M22. However, it does have one unique feature—if you look close, a "bar" of bright stars seems to stretch across its core.

For context, Antares is about 500 light years away, whereas M4 is at least ten times more distant. This may seem far, but M4 is one of the closest globular clusters to Earth.

A second globular cluster, M80, lies nearby. However, I've found this more challenging to identify conclusively from the suburbs in a small telescope.

Observing Log

Date:	Time:	Location:
Sky Conditions (Seeing/Transparency):		
Instrument (Magnification/Aperture):		
Notes:		

Sketch

Object of Interest: The Spider Cluster (M4)
Dark Sky Requirement: Suburbs
Object Type: Globular cluster
Brightness: 5.6
Distance: 7,200 light years
Apparent Size: 36 arcminutes

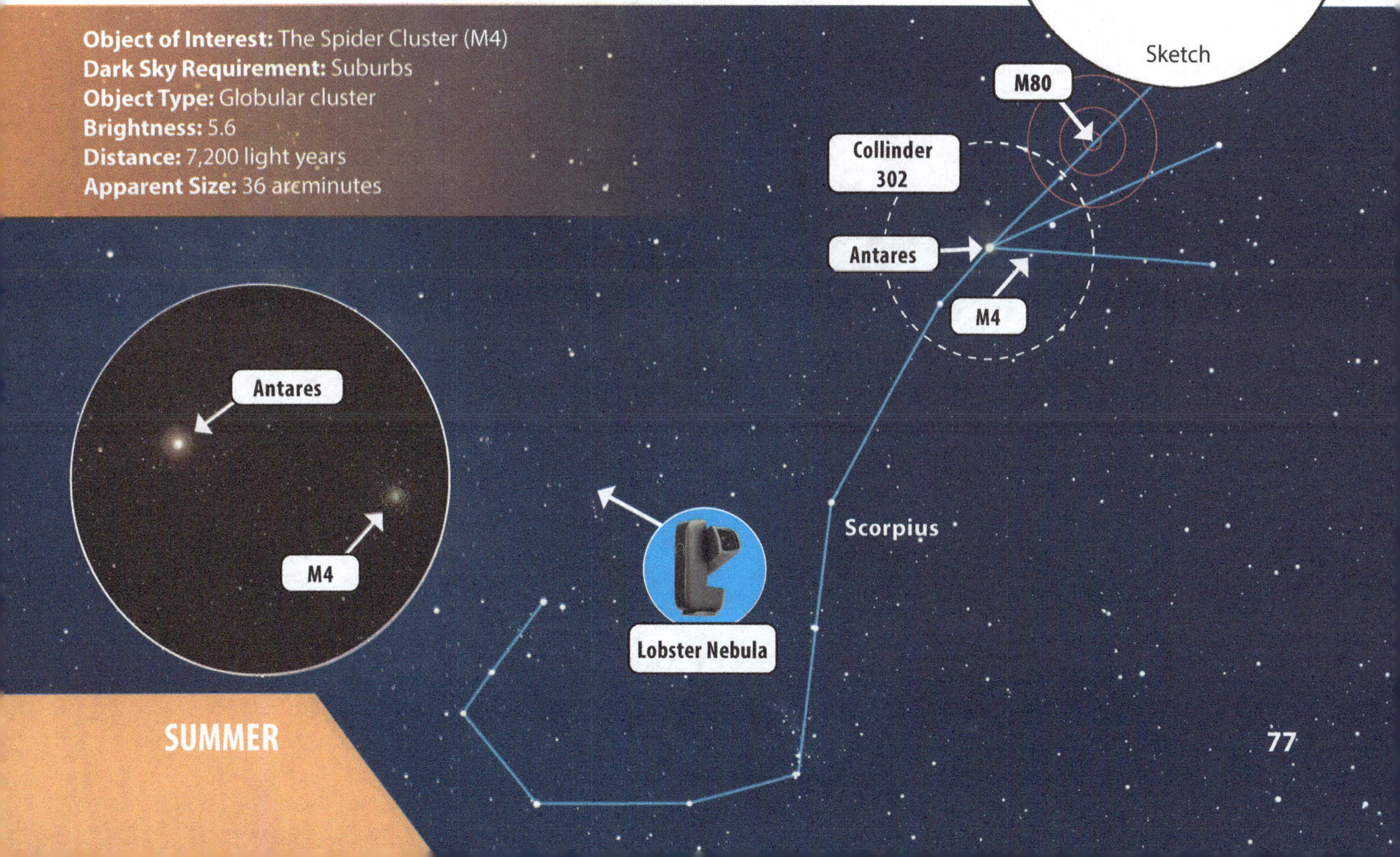

The Lagoon Nebula (M8)

From my closest national park, I have no trouble seeing the Lagoon Nebula without a telescope at all. It appears as an especially bright patch of sky against the background Milky Way. In the telescope, once your eyes have adapted to the dark, its detail is second only to winter's Orion Nebula.

M8 is a massive star-forming region where hot young stars illuminate the surrounding gas. It gets its name from the dark channels (formed by dust lanes) that appear when the nebula is viewed with a small telescope.

Within the nebula itself is an open star cluster, NGC 6530, which is nicknamed the "Hourglass Nebula."

Cameras, however, tend to pick up a significant amount of pink coloration due to the nebula's glowing hydrogen gas.

Observing Log

Date:	Time:	Location:

Sky Conditions (Seeing/Transparency):

Instrument (Magnification/Aperture):

Notes:

Sketch

Object of interest: Lagoon Nebula (M8)
Dark sky requirement: Suburbs
Object type: star-forming nebula
Brightness: 6
Distance: 4,300 light years
Apparent size: 90 × 40 arcminutes

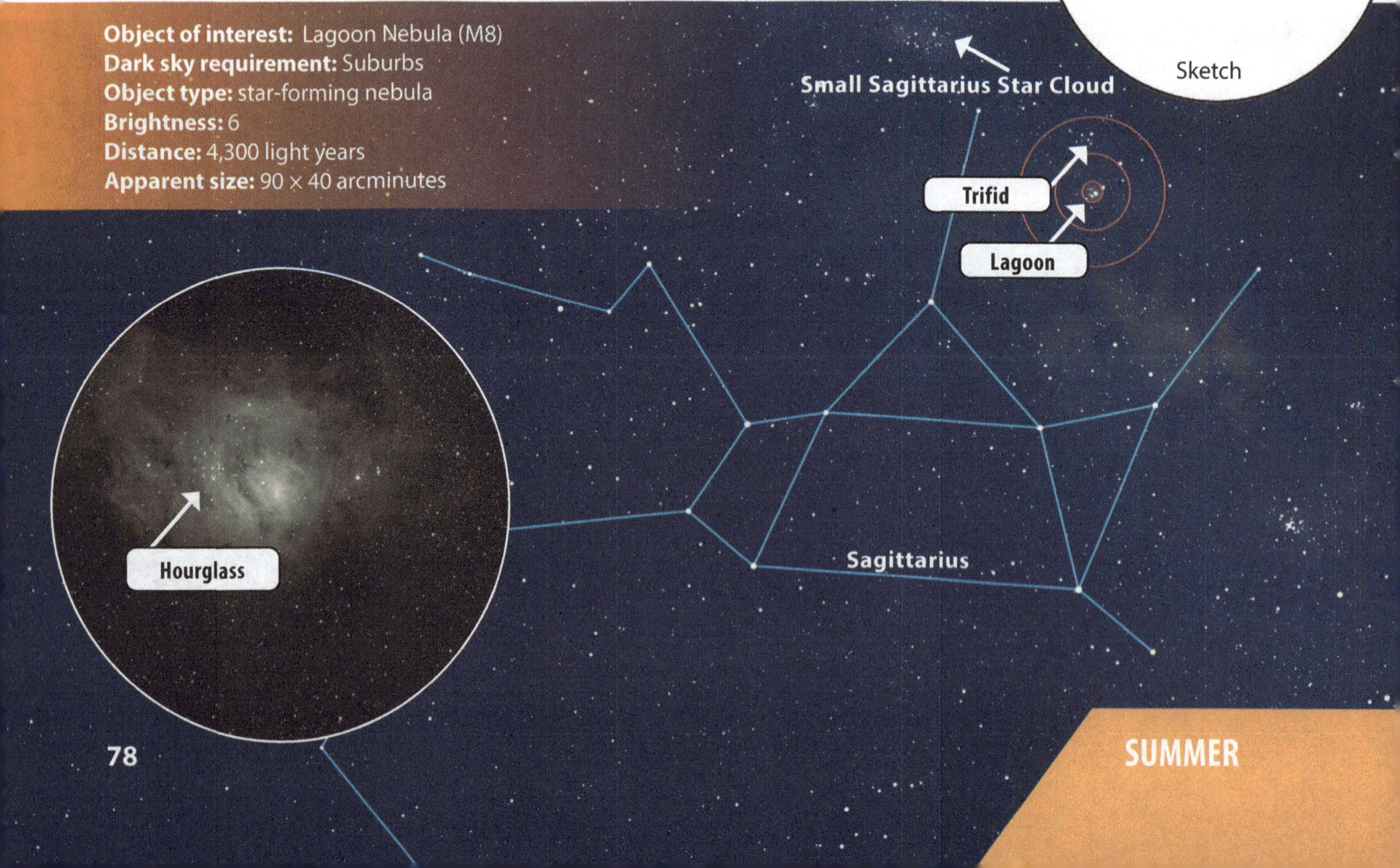

Great Sagittarius Cluster (M22)

From dark skies, I have little trouble seeing M22 as a small circular patch of light with no telescope at all. This, of course, assumes I haven't looked at my cell phone in about 30 minutes. M22 is slightly less well known than M13 (the Great Globular Cluster in Hercules) primarily because it's very low in the sky for most observers in the Northern Hemisphere.

However, astronomy communities have concluded that M22 is, by far, a much more impressive target. (Note that the southern hemisphere globular Omega Centauri and 47 Tucanae are brighter still.)

M22 contains an impressive five hundred thousand stars. Analyses with larger (and space-based) telescopes have confirmed that this cluster also contains two black holes and even a planetary nebula.

Observing Log

Date:	Time:	Location:
Sky Conditions (Seeing/Transparency):		
Instrument (Magnification/Aperture):		
Notes:		

Sketch

Object of Interest: M22
Dark Sky Requirement: Suburbs
Object Type: Globular cluster
Brightness: 5.1
Distance: 10,000 light years
Apparent Size: 32 arcminutes

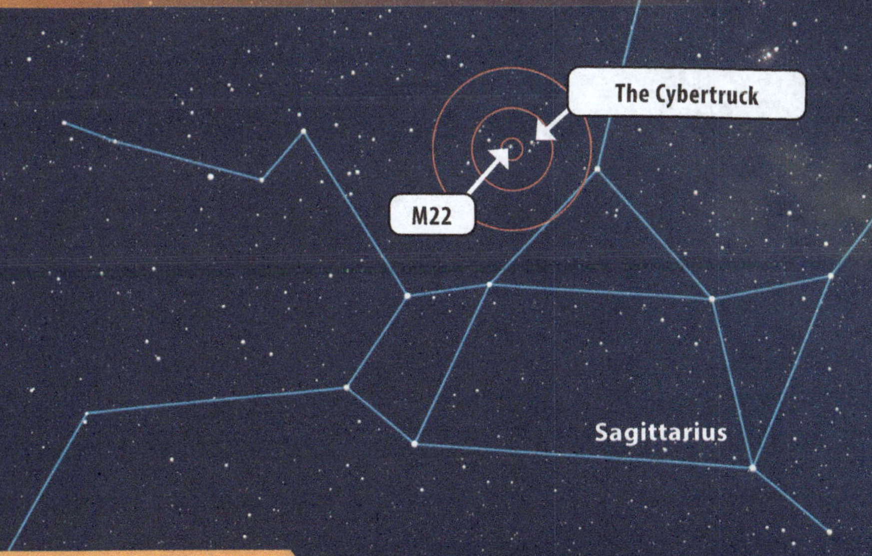

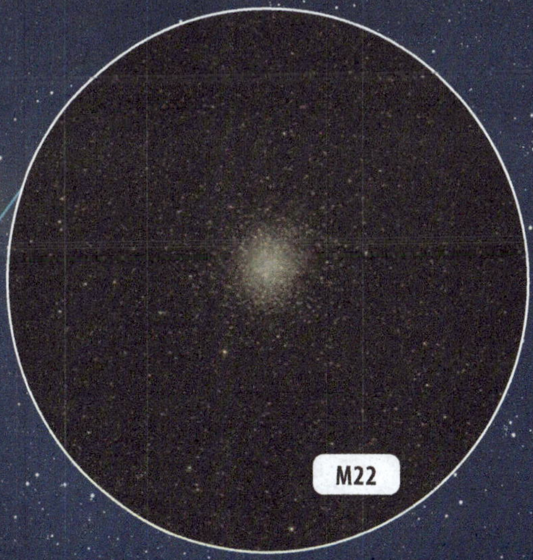

The Trifid Nebula (M20)

If you explore the area around the Teapot asterism, even without using a star map, you're bound to run into several interesting targets, including M20. At extremely low magnification, such as 20×, both the Trifid Nebula and the Lagoon Nebula can be seen within the same field of view.

If you look carefully, the smudge (remember, most nebulae look like beautiful smudges in a small telescope) is subdivided into three separate lobes. However, a fourth lobe is visible when using a camera.

You may also be able to resolve several small clusters of recently formed stars. These include an open cluster nicknamed "Webb's Cross," although the "cross" extends all the way back to M20 making this, according to astronomer Stephen O'Meara, technically an asterism.

Observing Log

Date:	Time:	Location:

Sky Conditions (Seeing/Transparency):

Instrument (Magnification/Aperture):

Notes:

Sketch

Object of Interest: The Trifid Nebula (M20)
Dark Sky Requirement: Suburbs
Object Type: Star-forming nebula
Brightness: 6.3
Distance: 5,200 light years
Apparent Size: 29 × 27 arcminutes

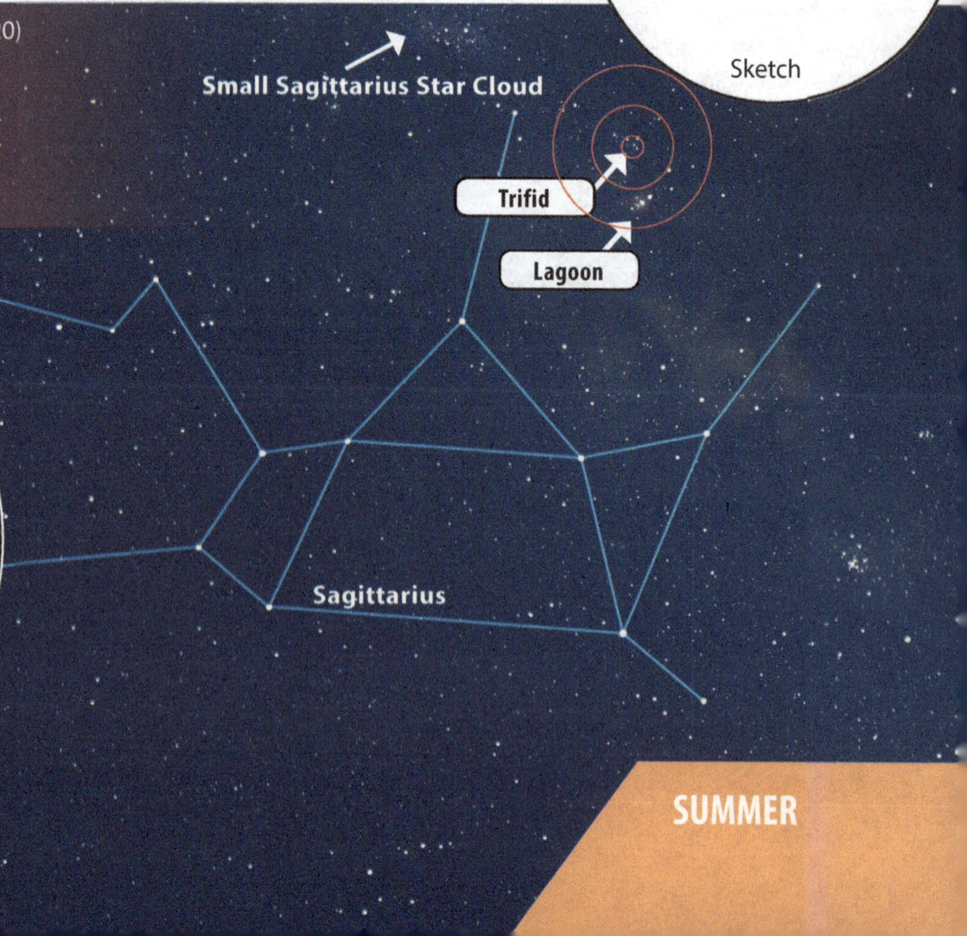

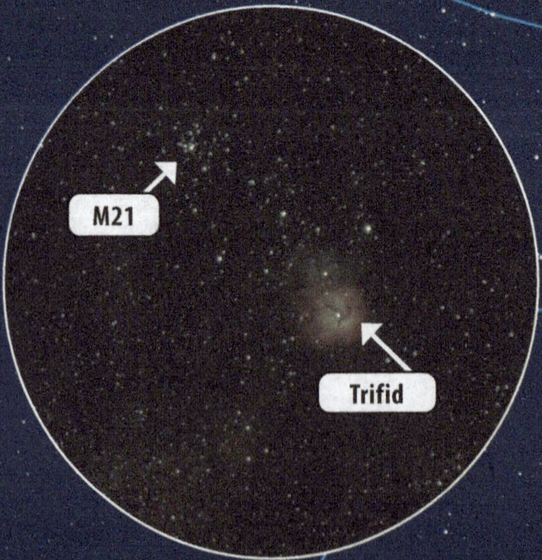

Ptolemy's Cluster (M7)

From my observing site in Eastern Canada, the large open cluster M7 barely makes it over the horizon. But from my closest national park, it can be seen easily without a telescope as a bright patch of light near the heart of the Milky Way. To take the photo for the "telescope view," I had to position the telescope high on a rock outcropping at Stargaze Nova Scotia.

M7 lies between the Teapot and Fishhook asterisms. The brightest stars seem to form an hourglass shape, with a tiny square-shaped gap in the middle. This cluster is also called Ptolemy's Cluster, named for the Greek astronomer who first recorded his observation over 1,000 years prior to the invention of the telescope.

Observing Log

Date:	Time:	Location:

Sky Conditions (Seeing/Transparency):

Instrument (Magnification/Aperture):

Notes:

Sketch

Object of Interest: Ptolemy's Cluster (M7)
Dark Sky Requirement: Suburbs
Object Type: Open cluster
Brightness: 3.2
Distance: 980 light years
Apparent Size: 80 arcminutes

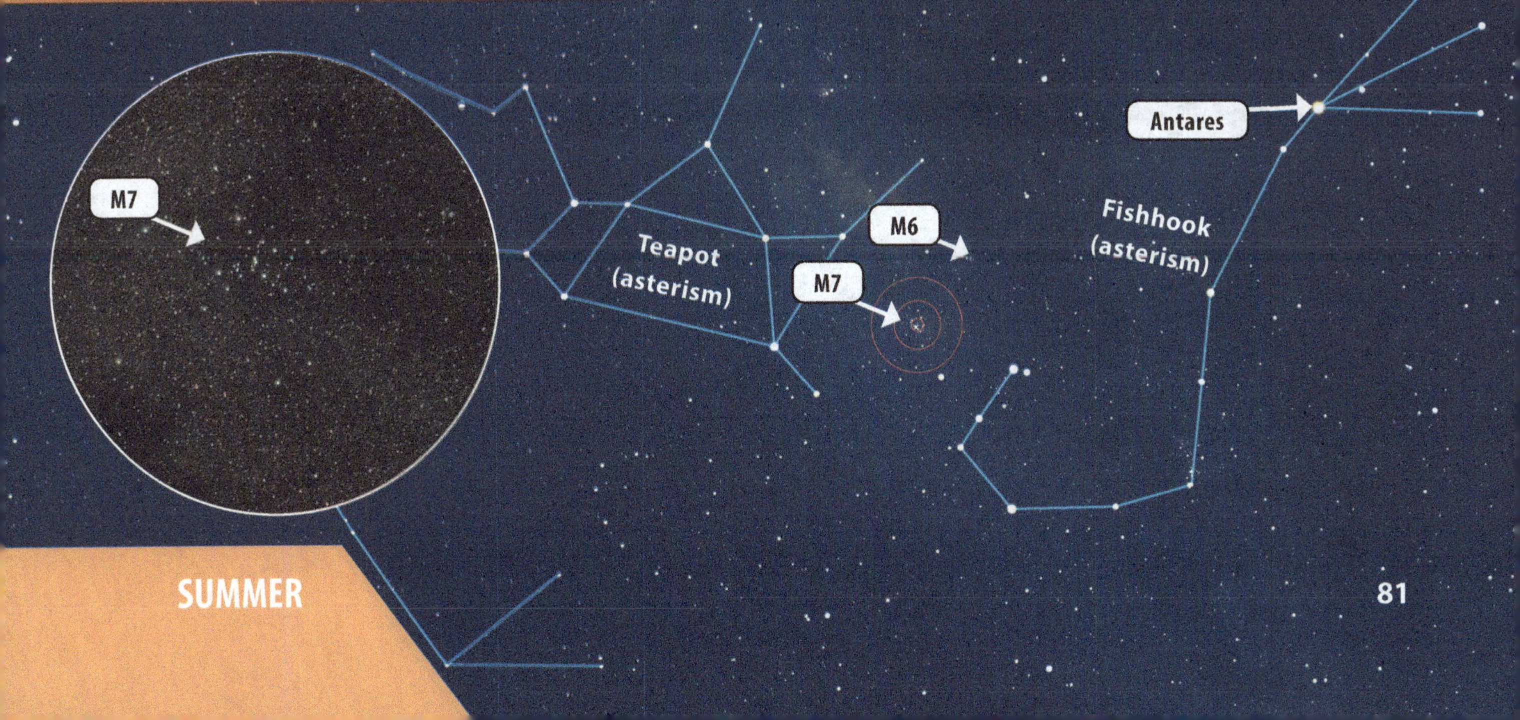

The Eagle Nebula (M16)

The Eagle Nebula is the location of the "Pillars of Creation," one of the most famous Hubble Space Telescope images of all time. This nebula is so intricate when imaged with a camera that one of the challenges with creating the "telescope view" image below was reducing the amount of detail. That said, astronomer and author Stephen O'Meara describes this as follows "tidal waves of dark matter that appear to be scrubbing away the bright coast of gas with their heavy ebb and flow."

It's possible to view the "pillars" with just your eyes in a large amateur-grade telescope, although you need ideal conditions. You'll also need to adjust the magnification to get the contrast just right. Remember, there is a specific level of magnification unique to your eyes where the exit pupil (the column of light leaving the telescope) matches your eye's pupil size, maximizing the quality of the view.

Observing Log

Date:

Time:

Location:

Sky Conditions (Seeing/Transparency):

Instrument (Magnification/Aperture):

Notes:

Sketch

Object of Interest: M16
Dark Sky Requirement: Suburb-visible
Object Type: Emission nebula
Brightness: 6.4
Distance: 5,700 light years
Apparent Size: ~30 arcminutes

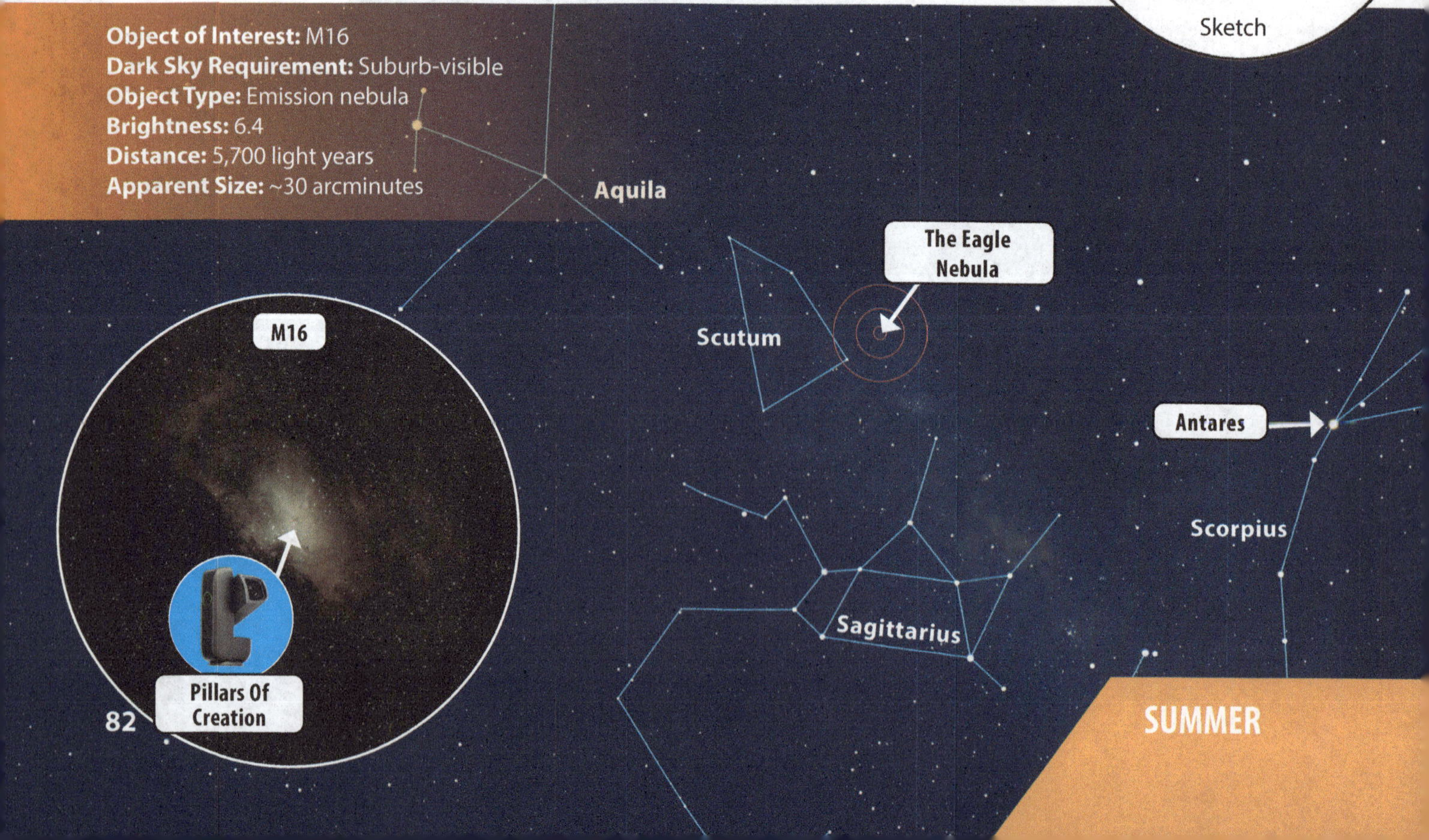

The Butterfly Cluster (M6)

The Butterfly Cluster is one of the "cutest" clusters in the sky (up there with the Swimming Alligator in Cepheus). Two colorful stars make up the "eyes" of the butterfly, while its wings and antennae are quite well defined. Like M7, this cluster is bright enough to see with the unaided eye, and in binoculars the two clusters fit into the same field of view.

At about 50×, the cluster fills the telescope's field of view, and dozens or even hundreds of stars will be visible depending on your aperture. Be sure to observe the star BM Scorpii (one of the eyes), noting its strikingly bright orange color.

To find M6, simply follow a line formed by the spout of the teapot. Alternatively, pan up from M7.

Observing Log

Date:	Time:	Location:
Sky Conditions (Seeing/Transparency):		
Instrument (Magnification/Aperture):		
Notes:		

Sketch

Object of Interest: The Butterfly Cluster M6
Dark Sky Requirement: Suburbs
Object Type: Open cluster
Brightness: 4.2
Distance: 1,600 light years
Apparent Size: 20 arcminutes

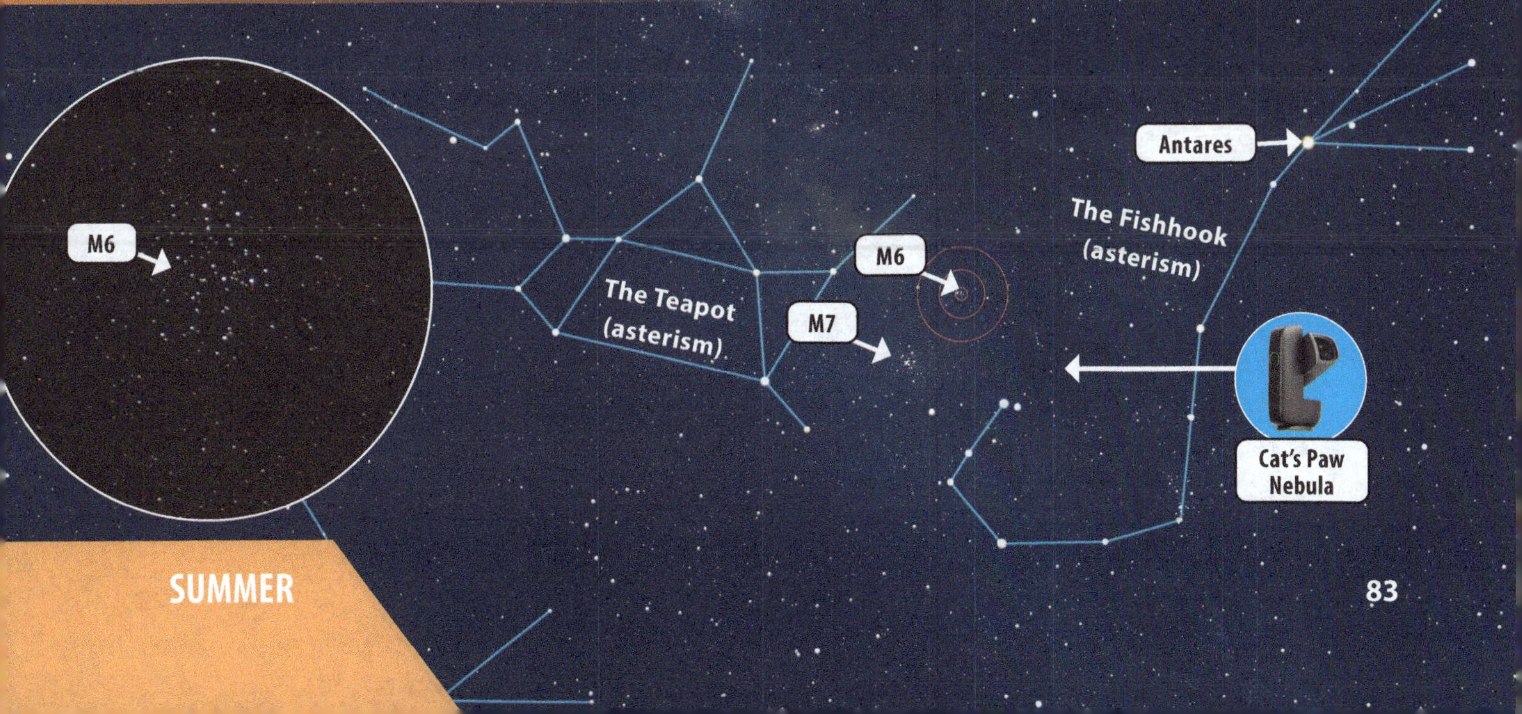

The Swan Nebula (M17)

M17 is another "star factory" where gas and dust swirl together to form young stars. This nebula is relatively compact compared to the Lagoon, but it is second only to the Orion Nebula in brightness.

The nebula clearly looks like a swan in photographs, but visually, at first glance, it appears as a rectangular smudge. When viewed with small telescopes, it is said to appear as the Greek letter omega (Ω) or even a horseshoe, according to astronomer Stephen O'Meara.

M17 also goes by a variety of names. The NGC catalog mentions its likeness to the number "2." SkySafari calls it the "Check-mark" and Wikipedia calls it the Lobster, though I was unable to trace the source of that last one.

Observing Log

Date:	Time:	Location:
Sky Conditions (Seeing/Transparency):		
Instrument (Magnification/Aperture):		
Notes:		

Sketch

Object of Interest: The Swan Nebula (M17)
Dark Sky Requirement: Suburb-visible
Object Type: Emission nebula
Brightness: 6.0
Distance: 4,200 light years
Apparent Size: 46 × 37 arcminutes

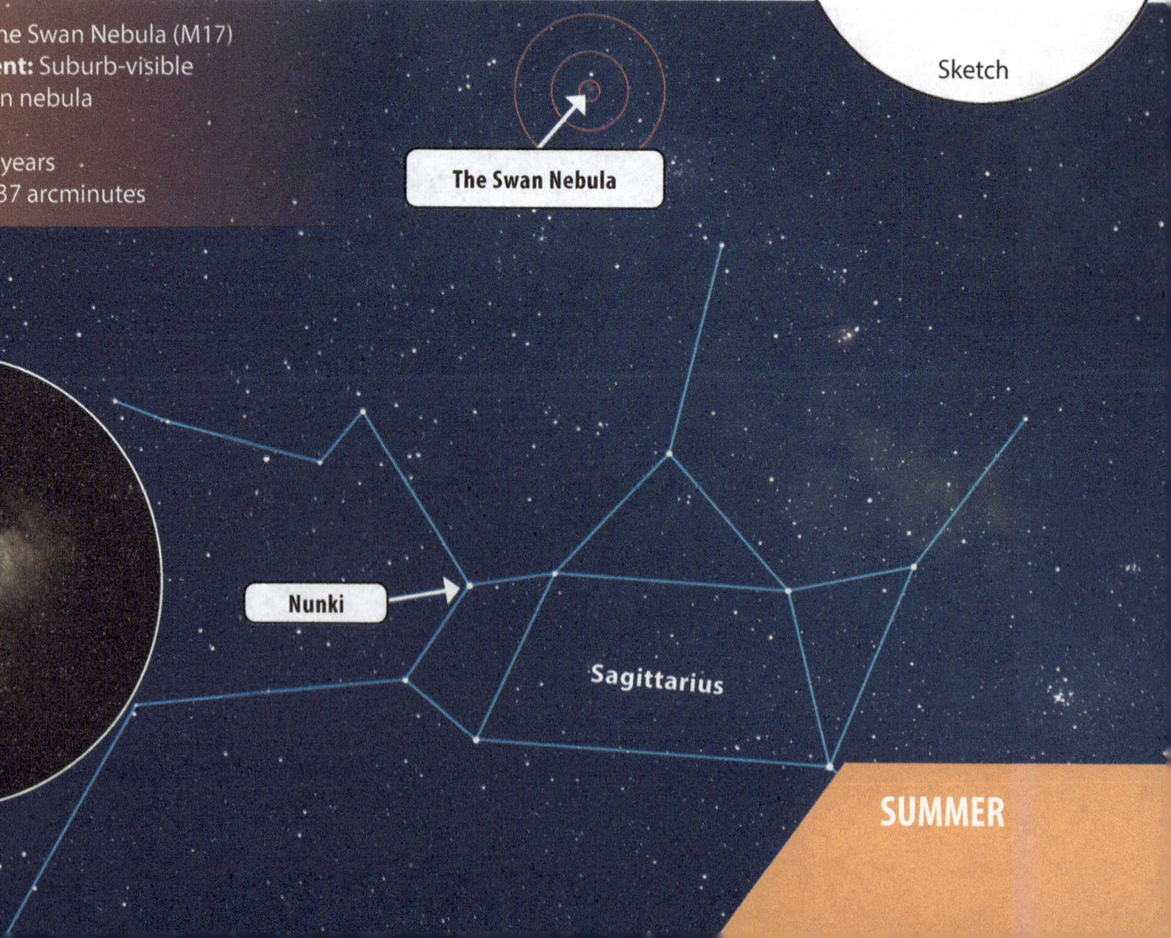

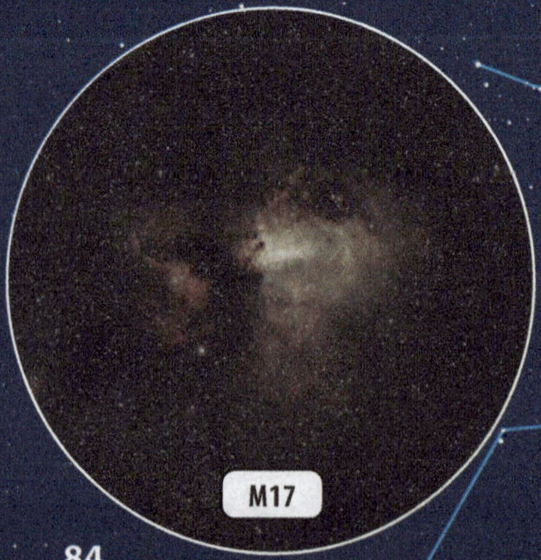

Albireo

Albireo is one of my favorite double stars. I'm sure to showcase it at pretty much every stargazing event due to the striking color contrast between the stars. The brighter of the two stars, Albireo A (Beta Cygni A), appears yellow or amber, while Albireo B (Beta Cygni B) is blue.

Although Albireo spends most of the winter below the horizon, there is a slightly more challenging to find almost identical double star, dubbed "Winter Albireo," (145 CMa) in Canis Major, which we'll explore on page 114.

This double star is found at the base of the Northern Cross. If you have a cell phone adapter for your telescope, this is a great star to photograph, as the phone's camera will accentuate the contrasting colors between the two stars.

Observing Log

Date:	Time:	Location:
Sky Conditions (Seeing/Transparency):		
Instrument (Magnification/Aperture):		
Notes:		

Sketch

Object of Interest: Albireo
Dark Sky Requirement: City-viewable
Object Type: Double star
Brightness: 3.1 and 4.7
Distance: 430 light years
Apparent Size: Point source(s)

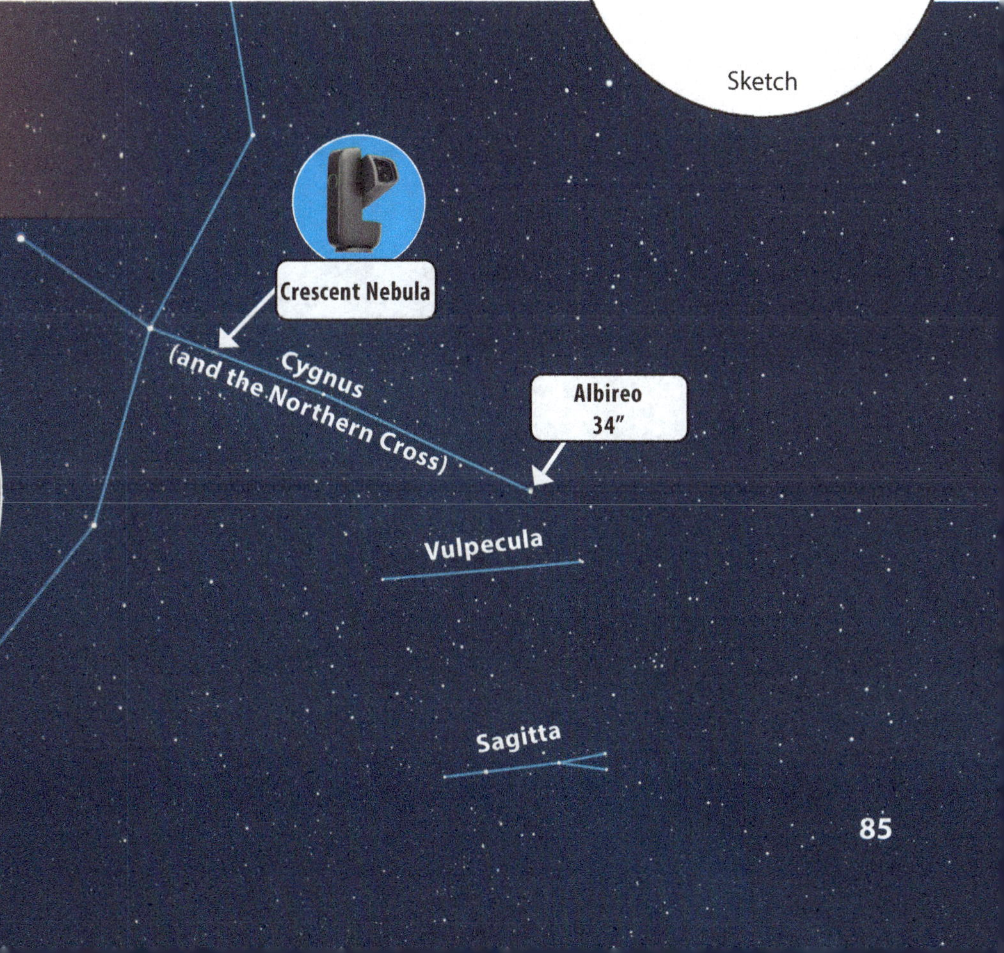

The Dumbbell (M27)

Discovered in 1764 by the French astronomer Charles Messier, the Dumbbell Nebula (M27) was the first planetary nebula ever discovered. It also has a very large apparent size in a telescope, at almost a quarter the diameter of the Full Moon.

M27 is located in the Summer Triangle, between the constellations Vulpecula and Sagitta. Interestingly, the Dumbbell Nebula wasn't given its name until 1833, when astronomer John Herschel made this record: "A nebula shaped like a dumb-bell, with the elliptic outline completed by a feeble nebulous light."

Most planetary nebulae look round, like … planets. But this nebula looks more like an apple core. It can be found by searching the area above Sagitta (the Arrow).

Observing Log

Date:	Time:	Location:
Sky Conditions (Seeing/Transparency):		
Instrument (Magnification/Aperture):		
Notes:		

Sketch

Object of Interest: The Dumbbell (M27)
Dark Sky Requirement: Suburbs
Object Type: Planetary nebula
Brightness: 7.1
Distance: 1,400 light years
Apparent Size: 8 × 6 arcminutes

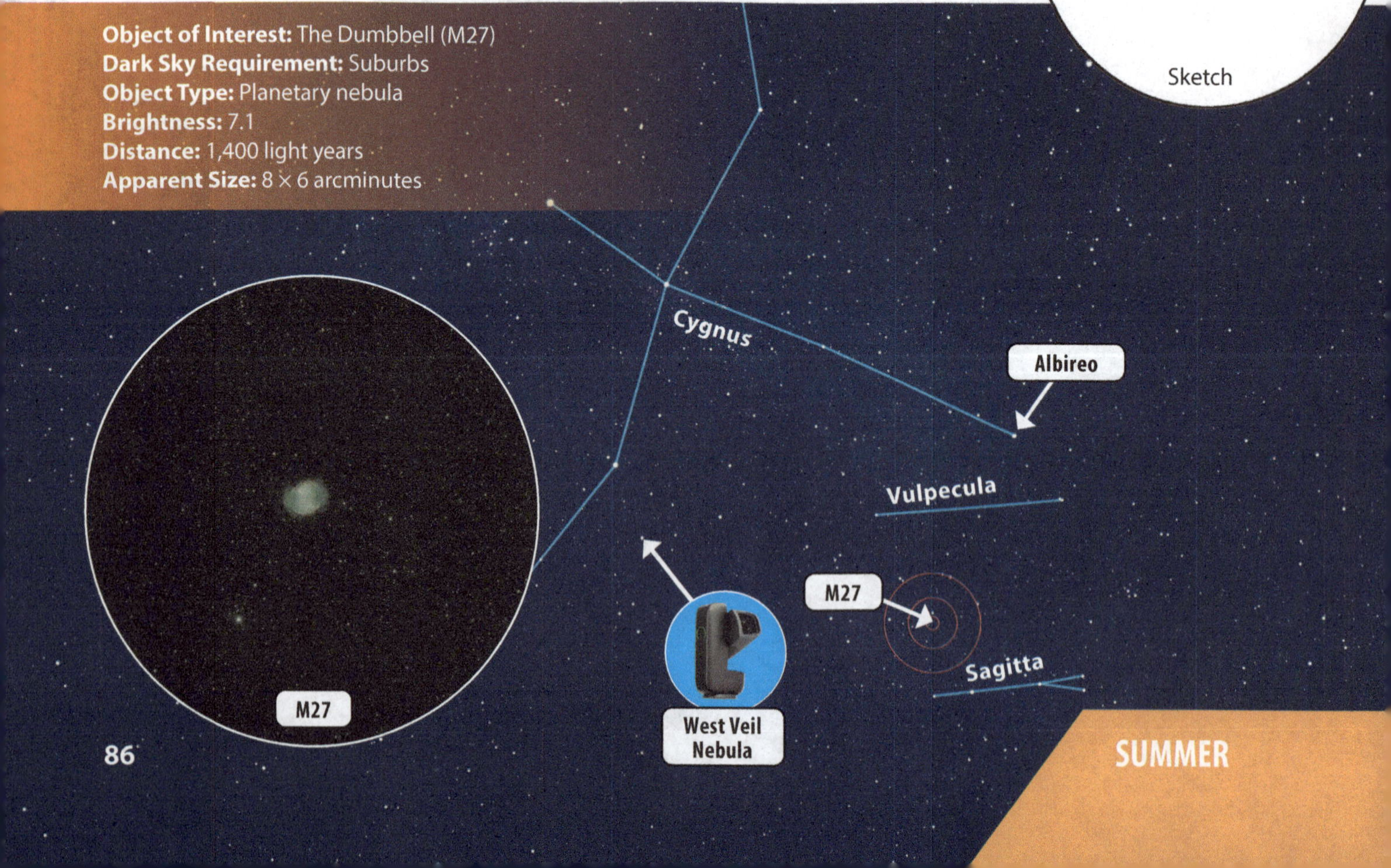

Great Globular Cluster (M13)

The Great Globular Cluster in Hercules, or M13, is one of the brightest deep-sky objects. On more than one occasion, I've run into it accidentally while randomly pointing the telescope at the sky. It's one of the largest globular clusters, containing several hundred thousand stars.

M13 is also insanely densely populated. There are only two star systems within five light years of Earth, but if we lived near the core of M13, there would be literally tens of thousands of star systems within that same volume of space.

To find M13, search the side of the Keystone (an asterism in Hercules); pan around the edge of the square until you find it. In the summer, M13 may be so high in the sky the telescope may need to point straight up, which is challenging for many beginner's telescope designs.

Observing Log

Date:	Time:	Location:
Sky Conditions (Seeing/Transparency):		
Instrument (Magnification/Aperture):		
Notes:		

Sketch

Object of Interest: The Great Cluster in Hercules (M13)
Dark Sky Requirement: Suburbs
Object Type: Globular cluster
Brightness: 5.8
Distance: 23,000 light years
Apparent Size: 20 arcminutes

SUMMER

The Wild Duck Cluster (M11)

The Wild Duck Cluster (M11) is a compact, open star cluster that looks almost like a globular cluster. The easiest way to find it in your telescope is to follow the tail of the Eagle in the constellation Aquila along a curved trail of stars at its base. The cluster will appear almost like a globular cluster, but it's not. Although the cluster contains thousands of stars, only the brightest pop out when viewed through a backyard telescope. **Note:** Technically, M11 resides within the borders of the constellation Scutum (the Shield).

M11 was originally named the Wild Duck Cluster by a British Naval officer named William Henry Smyth. But some call it the "Borg Cube" due to its resemblance to a fictional spacecraft from the *Star Trek Universe*.

Observing Log

Date:	Time:	Location:
Sky Conditions (Seeing/Transparency):		
Instrument (Magnification/Aperture):		
Notes:		

Sketch

Object of Interest: The Wild Duck Cluster (M11)
Dark Sky Requirement: Suburbs
Object Type: Open cluster
Brightness: 5.8
Distance: 6,100 light years
Apparent Size: 32 arcminutes

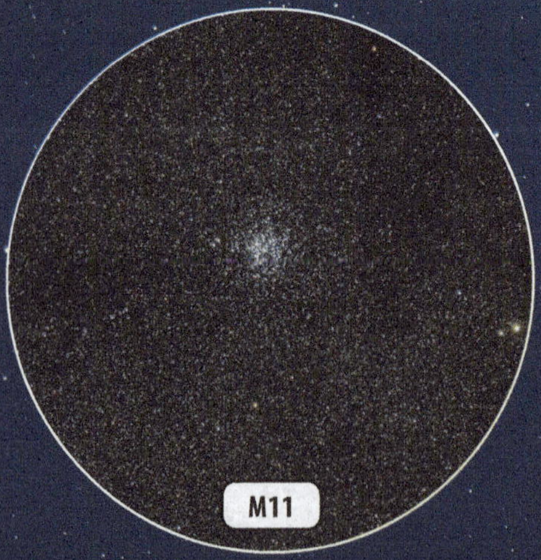

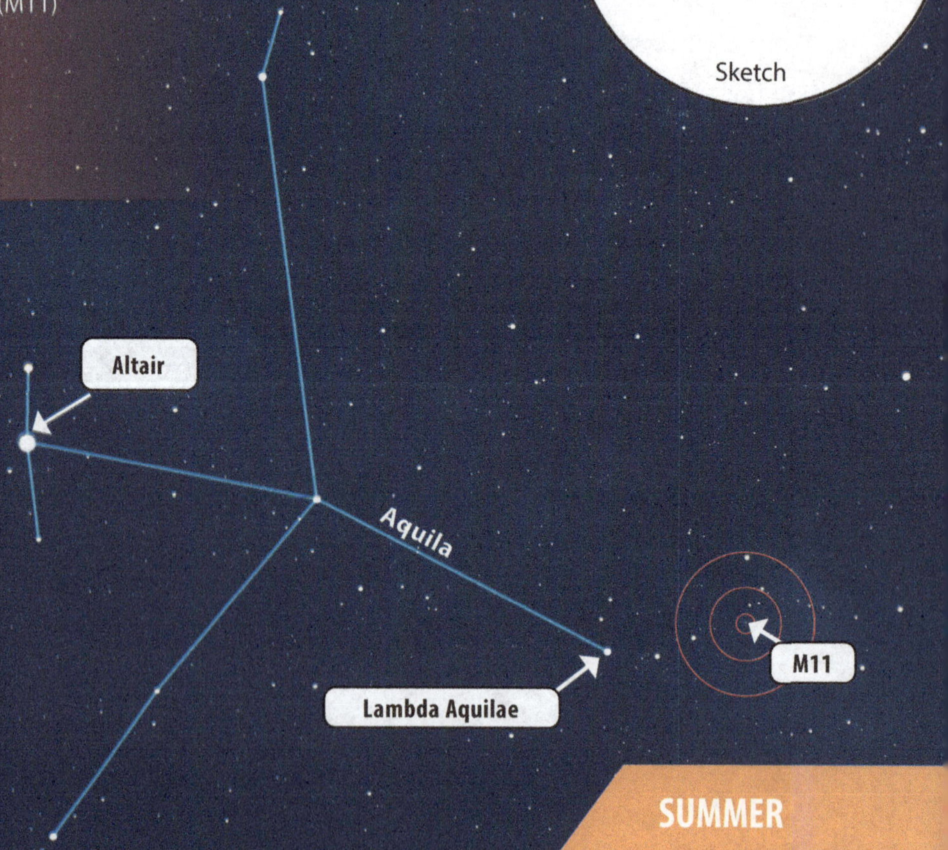

The Ring Nebula (M57)

The Ring Nebula appears about as large as Jupiter through a telescope, but not nearly as bright. The challenge, in a small telescope, is to clearly make out the hole in the ring. In order to view the center of the ring, you may need a telescope with a lens or mirror at least 100 mm in diameter, but M57's disk shape is visible in smaller telescopes.

This nebula was formed when a red giant star shed its outer shell of ionized gas, leaving only a white dwarf star where the red giant once was.

To find the Ring Nebula, pan the telescope between the two stars opposite Vega in the diamond pattern in the constellation Lyra.

Observing Log

Date:	Time:	Location:

Sky Conditions (Seeing/Transparency):

Instrument (Magnification/Aperture):

Notes:

Sketch

Object of Interest: The Ring Nebula (M57)
Dark Sky Requirement: Suburbs
Object Type: Planetary nebula
Brightness: 5.8
Distance: 1,400 light years
Apparent Size: ~1.3 arcminutes

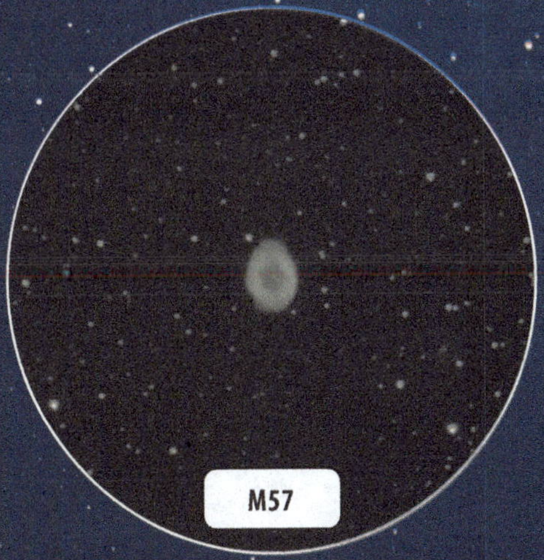

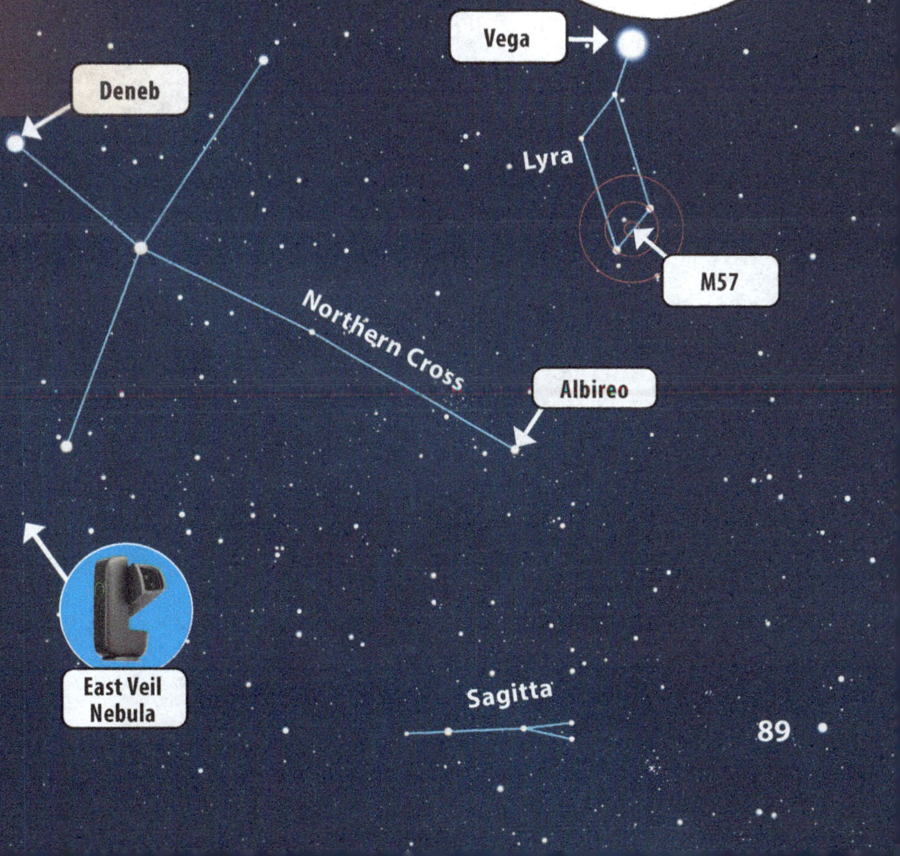

Coathanger (Brocchi's Cluster)

Although the Coathanger is one of the most popular binocular targets, it's a beautiful target for small telescopes at lower magnifications. This asterism of ten stars is slightly too dim to see with the unaided eye from the suburbs. Its stars, mostly fifth and sixth magnitude in brightness, are just out of visual range under most seeing conditions.

To find the Coathanger, first identify Albireo at the foot of the Northern Cross, then Sagitta, the Arrow. The Coathanger is found in between and slightly south of these objects.

The Coathanger's formal designation is Collinder 399, but was popularized by a draftsman named Dalmero Brocchi (1871–1955), who the American Astronomical Society nicknamed the "The Wizard of Puget Sound" for his contributions in the field of variable star observing and public outreach.

Observing Log

Date:	Time:	Location:
Sky Conditions (Seeing/Transparency):		
Instrument (Magnification/Aperture):		
Notes:		

Sketch

Object of Interest: The Coathanger
Dark Sky Requirement: Suburbs
Object Type: Asterism
Brightness: 3.4
Distance: 4,200 light years
Apparent Size: 89 arcminutes

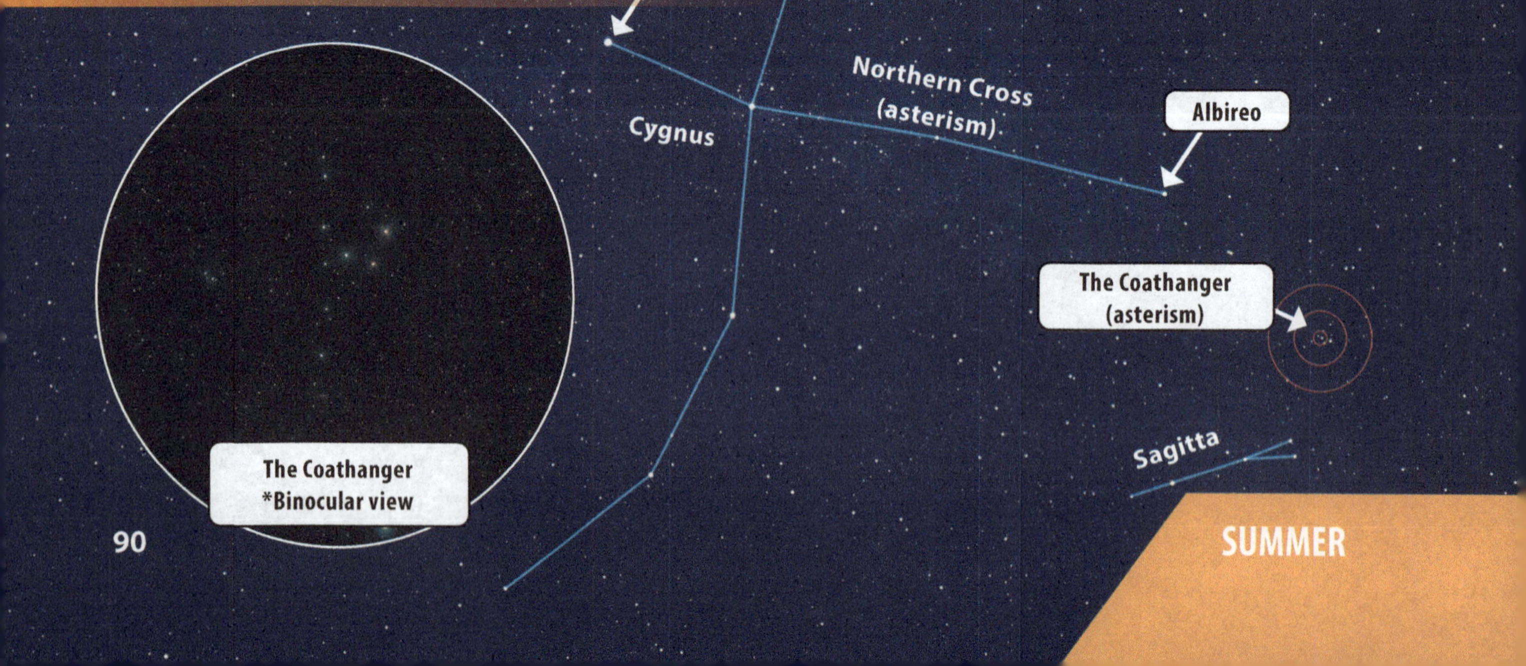

The Arrow Cluster (M71)

Sagitta, or The Arrow Cluster, looks like Harry Potter's broomstick zipping across the sky. Near the center of the Arrow, like a heart pierced by Cupid, is M71, a sparse globular cluster.

This star cluster is only twelve thousand light-years away, which is almost twice as close as the famous Great Globular Cluster in Hercules, M13.

Several of the cluster's individual stars are bright enough to be seen in light-polluted skies, making it an acceptable target for urban star parties.

It is not very dense, and almost shares its appearance with the open cluster M11. Due to its low density and small number of stars, M71 is unofficially referred to as a "loose globular," a designation it shares with the springtime cluster, M68.

Observing Log

Date:	Time:	Location:

Sky Conditions (Seeing/Transparency):

Instrument (Magnification/Aperture):

Notes:

Sketch

Object of Interest: The Arrow Cluster (M71)
Dark Sky Requirement: Suburbs
Object Type: Loose globular cluster
Brightness: 8.2
Distance: 13,000 light years
Apparent Size: 3.3 arcminutes

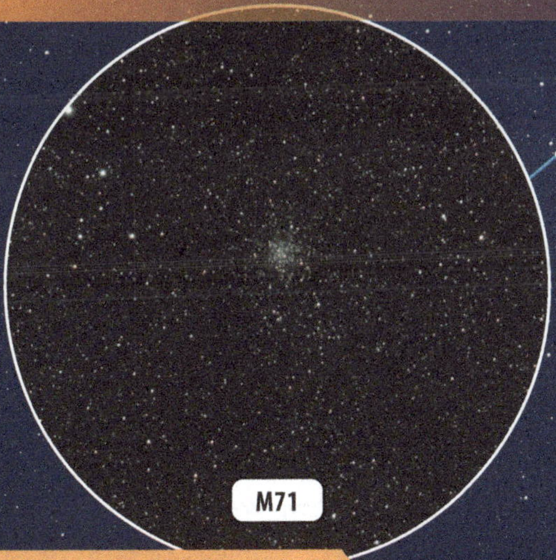

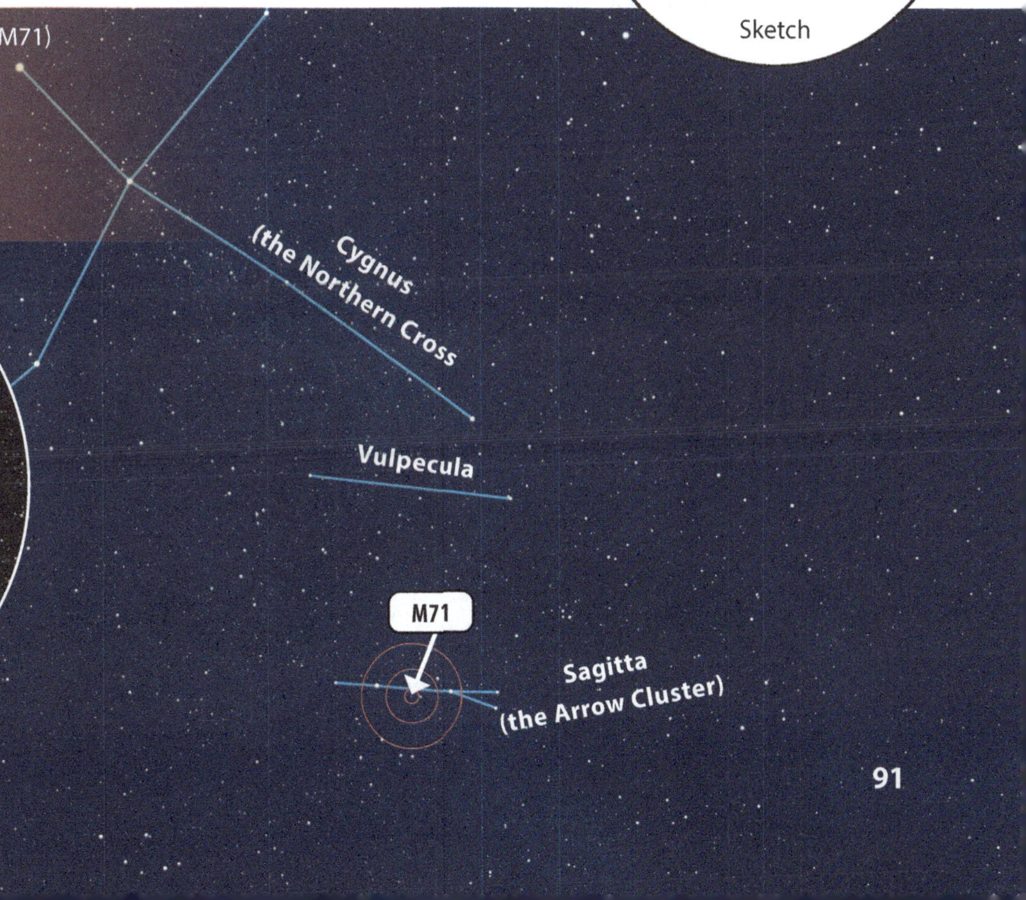

SUMMER

Summer Double Stars & Asterisms

The summer sky is brimming with so many deep sky wonders that observing asterisms and minor double stars is often an afterthought. While the finest asterism, the Coathanger, and the most famous double star, Albireo, have their own pages, there are several popular asterisms and double stars that I still consider showcase targets.

The Double Double is a double, double star, located right next to the bright star Vega.

61 Cygni has about the same separation as double star Albireo, except that both stars are gold in color!

Graffias is a double star that forms the left claw in Scorpius.

The "Hockey Stick" asterism in Cygnus is interesting for both its shape and the colors of its stars.

"Job's Coffin" is an asterism that forms the diamond in the Delphinus constellation—can you spot the "7" within it?

Observing Log

Date:	Time:	Location:

Sky Conditions (Seeing/Transparency):

Instrument (Magnification/Aperture):

Notes:

Sketch

Although not an asterism, if you can see the southernmost parts of Scorpius, the Northern Jewel Box—an open cluster—can be found near the base of the tail of the scorpion.

The Hockey Stick asterism

The "7" (seen when mirror reversed) in Job's Coffin

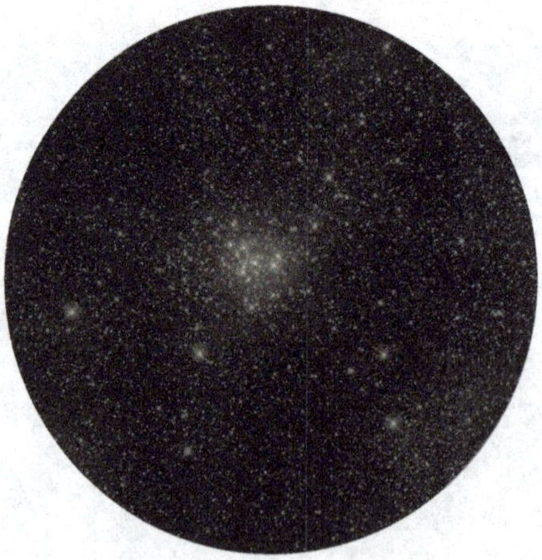

Northern Jewel Box (C76)

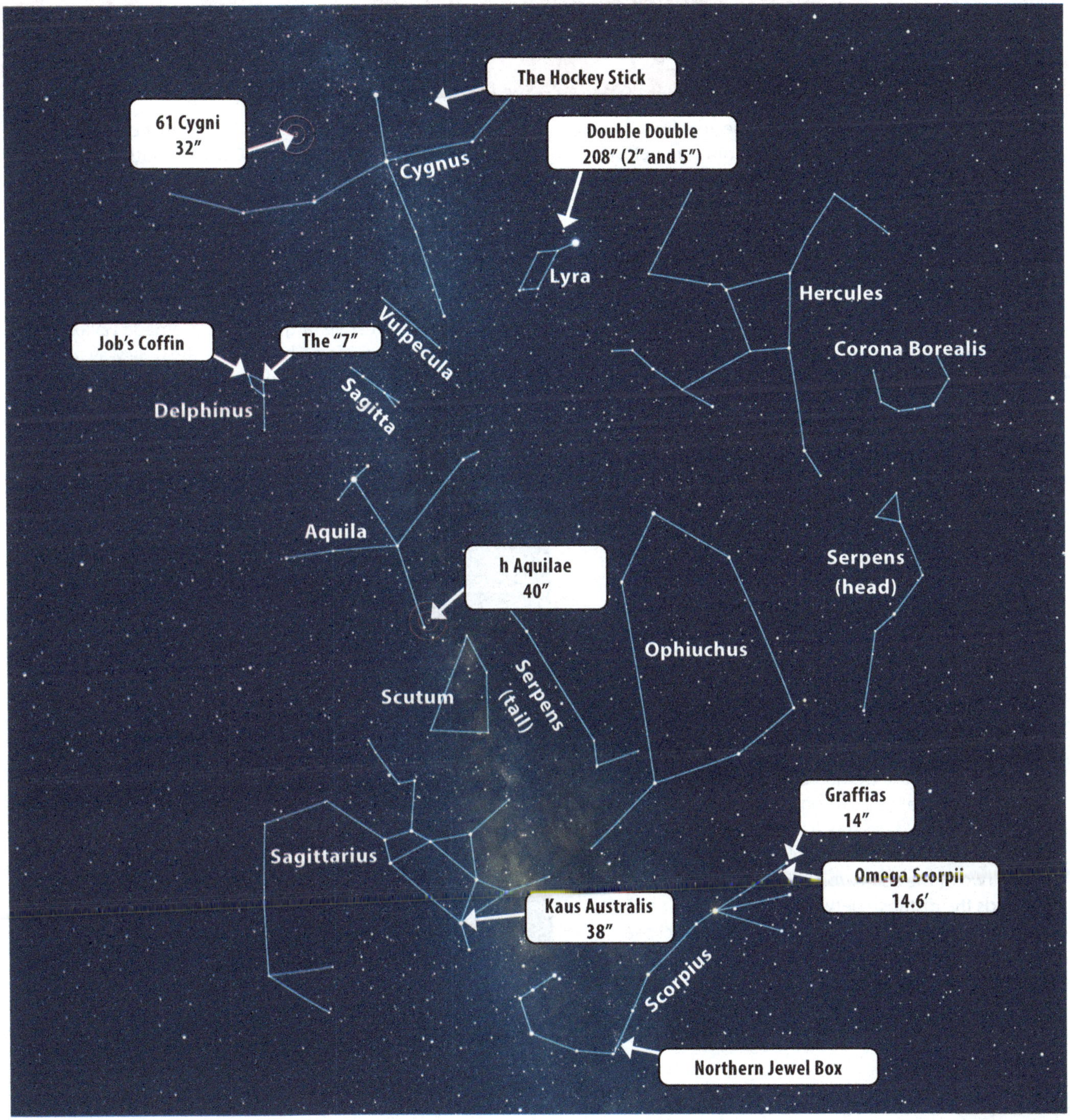
The Hockey Stick
61 Cygni
32″
Double Double
208″ (2″ and 5″)
Cygnus
Lyra
Hercules
Corona Borealis
Job's Coffin
The "7"
Vulpecula
Sagitta
Delphinus
Aquila
h Aquilae
40″
Serpens
(head)
Ophiuchus
Scutum
Serpens
(tail)
Graffias
14″
Sagittarius
Omega Scorpii
14.6′
Kaus Australis
38″
Scorpius
Northern Jewel Box

Autumn Targets

Autumn is the absolute best time of the year for stargazing for two reasons. First, the nights are getting longer and you don't have to wait long for it to get dark. Second, some of the best summer targets are still visible. At mid-northern latitudes, Sagittarius is visible until late October, while the Summer Triangle sticks around almost until the end of December!

Not only do you get the summer targets, but the best circumpolar targets, such as those near Cassiopeia, are also high in the sky.

The autumn sky is dominated by an asterism known as the Great Square of Pegasus. In light-polluted skies, when looking south, sometimes it appears that the autumn sky contains almost no stars at all! This is because Pisces, Cetus, Aquarius, and Capricornus contain few bright stars.

Most of the best autumn targets are found near the constellations Perseus and Andromeda.

The brightest star in Perseus is called Mirfak, but the second brightest star, Algol (Beta Persei), is considerably more interesting.

Algol is known as the Demon Star. Ancient cultures thought of this star as a symbol of bad luck or dark times. Every three days, this star dims for about 10 hours. This occurs because Algol is part of a multi-star system. Two of the stars orbit very close together (about two thirds the distance between the Sun and Earth) and eclipse each other as viewed from Earth. The eclipse lasts about 10 hours, and this event causes the drop in Algol's apparent brightness.

Mirfak is found at the center of one of the most beautiful clusters in the sky, the Alpha Persei Cluster. Another star, Mirach, is helpful in locating the Andromeda Galaxy.

The star Fomalhaut is the brightest star in the autumn sky. However, for Northern Hemisphere observers, it is often found very low on the southern horizon.

Autumn's most impressive target, the Andromeda Galaxy, is best viewed in binoculars. This is really the only way to capture the entire structure within a single field of view. It's generally considered a rule that the quality of the view through a telescope increases with aperture. However, in this case of the Andromeda Galaxy, and most large extended objects for that matter, the darkness of your sky and your eye's pupil size matter far more.

Autumn Sky at a Glance

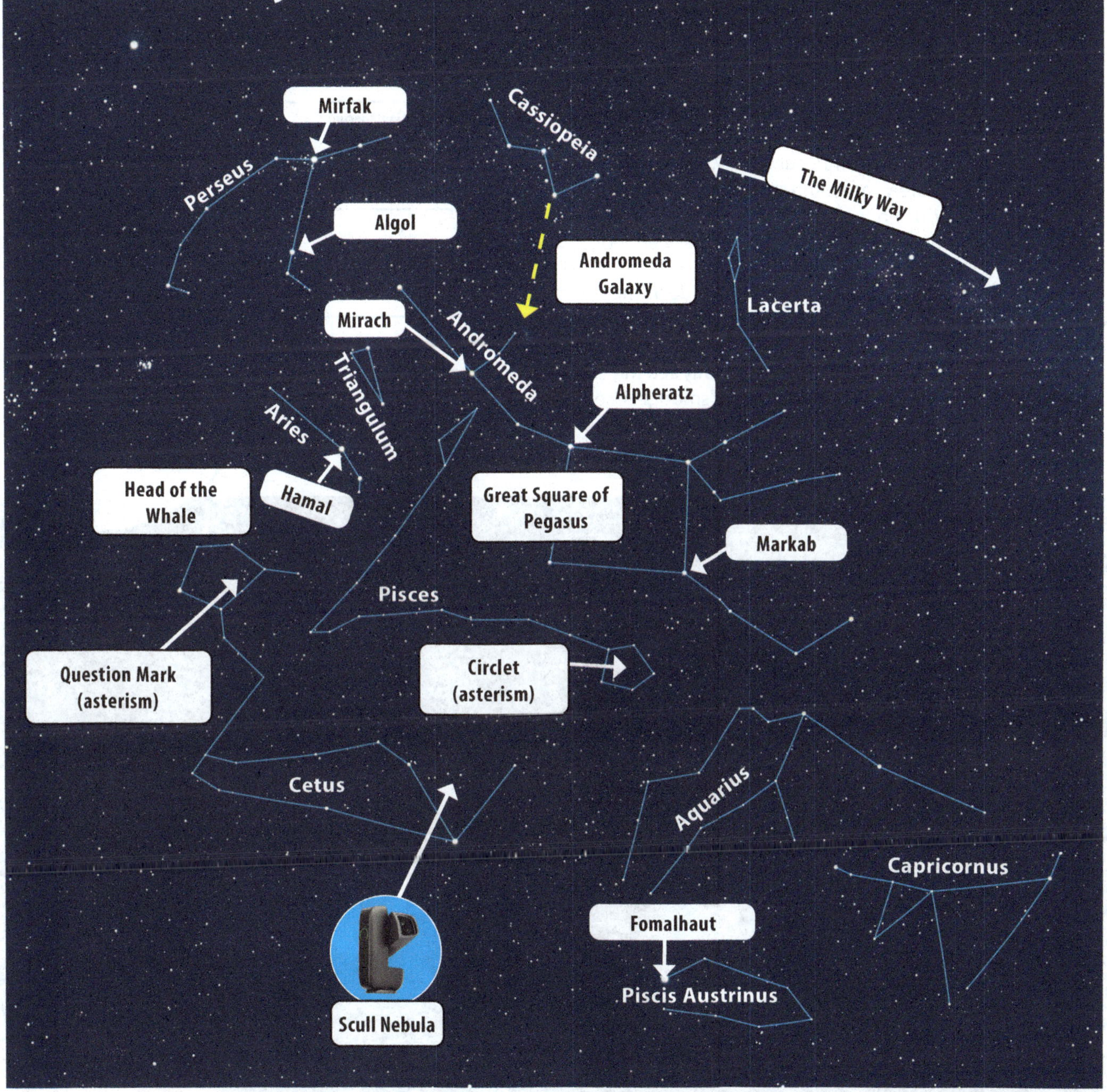

The Andromeda Galaxy (M31)

Before the twentieth century, the Milky Way was thought to be the only galaxy in the universe and the Andromeda Galaxy was called the Andromeda Nebula. Some astronomers dubbed objects that seemed to reside outside of our galaxy "Island Universes," while others believed these were nebulae that resided within our galaxy. It wasn't until Edwin Hubble accurately measured the distance to the Andromeda Galaxy, using the new 100-inch telescope at Mount Wilson Observatory, that the debate finally ended.

The Andromeda Galaxy appears six times as wide as the Full Moon! Therefore, binoculars, in dark skies, tend to be the best way to observe this object. When you see the Andromeda Galaxy in a telescope, you primarily see the galactic core, which appears as a bright gray smudge. Use your lowest magnification, and see if you can pick out the satellite galaxies as well: M32 located near the core, and M110 located northward of the core.

Observing Log

Date: | Time: | Location:

Sky Conditions (Seeing/Transparency):

Instrument (Magnification/Aperture):

Notes:

Sketch

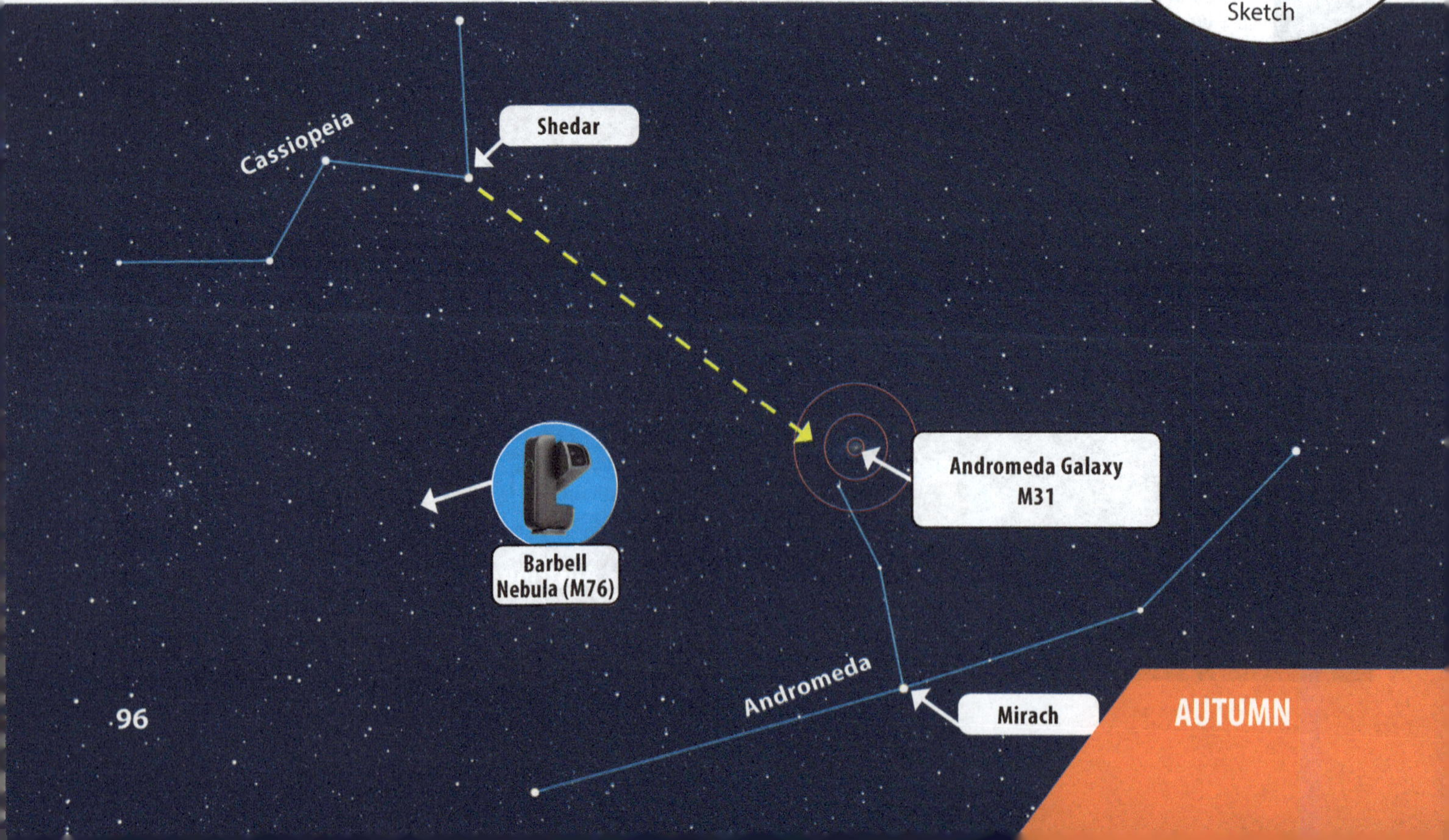

Object of Interest: The Andromeda Galaxy (M31)
Dark Sky Requirement: Suburbs
Object Type: Spiral galaxy
Brightness: 3.3
Distance: 2.5 million light years
Apparent Size: 178 × 70 arcminutes

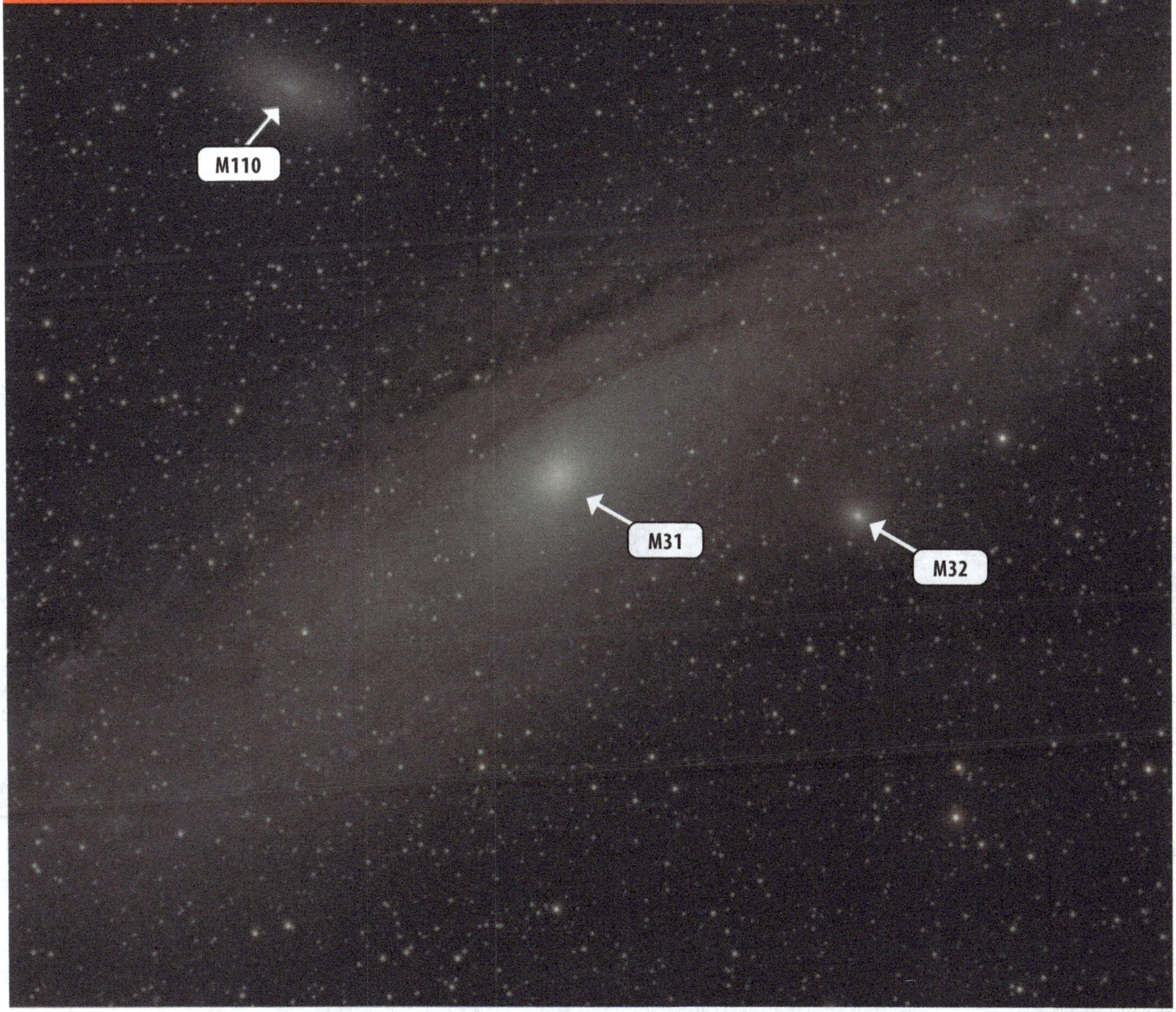

Triangulum Galaxy (M33)

The Triangulum Galaxy is a spiral galaxy located almost three million light years from the Milky Way. Astronomers have discovered a bridge of hydrogen and stars extending between M33 and M31, and have determined that these two galaxies collided in the distant past, and will most likely collide again in the future[1].

Start your search for M33 at the Andromeda Galaxy. Notice the two bright stars below it, and follow these stars down to the Triangulum Galaxy.

M33 is quite a bit dimmer than M31, and I've found it far more challenging to identify from the suburbs. From very dark skies, however, even in binoculars, this galaxy will pop out from the background.

Observing Log

Date:	Time:	Location:
Sky Conditions (Seeing/Transparency):		
Instrument (Magnification/Aperture):		
Notes:		

Sketch

Object of Interest: The Pinwheel/Triangulum Galaxy (M33)
Dark Sky Requirement: Dark skies
Object Type: Spiral galaxy
Brightness: 5.8
Distance: 2.8 million light years
Apparent Size: 62 × 37 arcminutes

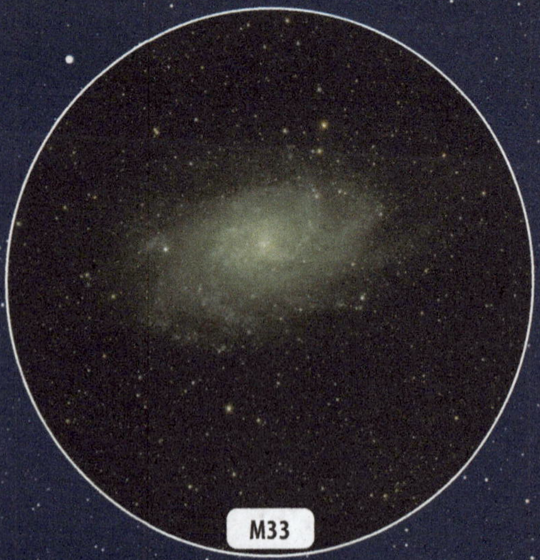

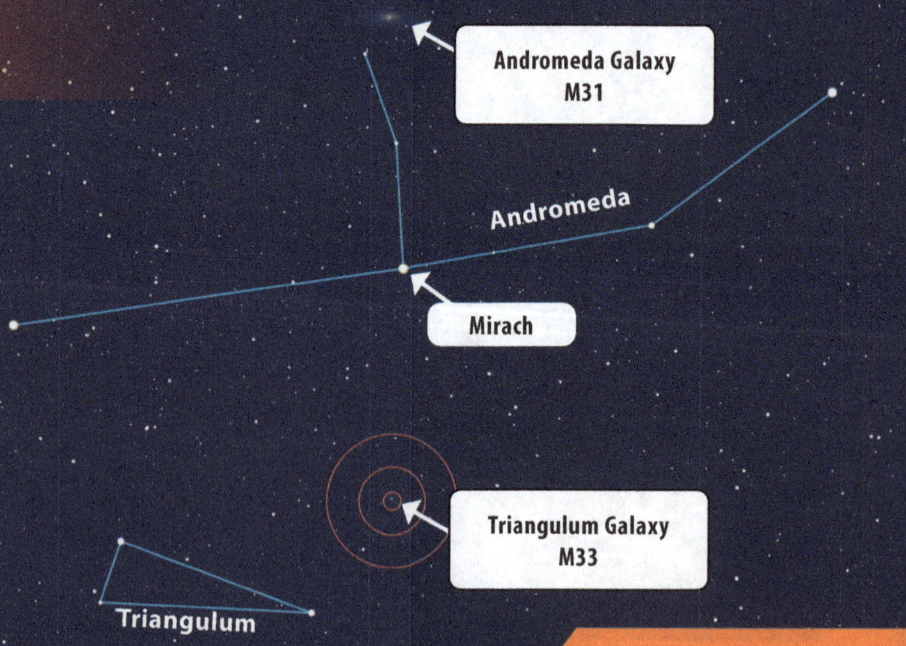

[1] The Astrophysical Journal, 751:74 (pp 26), 2012 May 20

Pegasus Clusters (M2 & M15)

While you may still be able to catch M13 and M22 before they set below the horizon, the autumn sky hosts two large globular clusters that flank the head of Pegasus.

From the Northern Hemisphere, Pegasus the Winged Horse appears upside down. A chain of four stars dangles off the great square, representing the horse's head and neck. M15 is found by following a line through the face of Pegasus, while M2 is found by following a line straight through the neck.

As a bit of a joke, a few years back I placed a $99, 100-mm Dobsonian (kid's telescope) on an EQ mount and took an image of M15. Here you can see the difference that aperture makes, as we compare the image of M2 taken with the Burke-Gaffney Observatory's Planewave CDK 24-inch telescope.

Observing Log

Date:	Time:	Location:

Sky Conditions (Seeing/Transparency):

Instrument (Magnification/Aperture):

Notes:

Sketch

Object of Interest: M2
Dark Sky Requirement: Dark skies
Object Type: Globular cluster
Brightness: 6.5
Distance: 38,000 light years
Apparent Size: 16 arcminutes

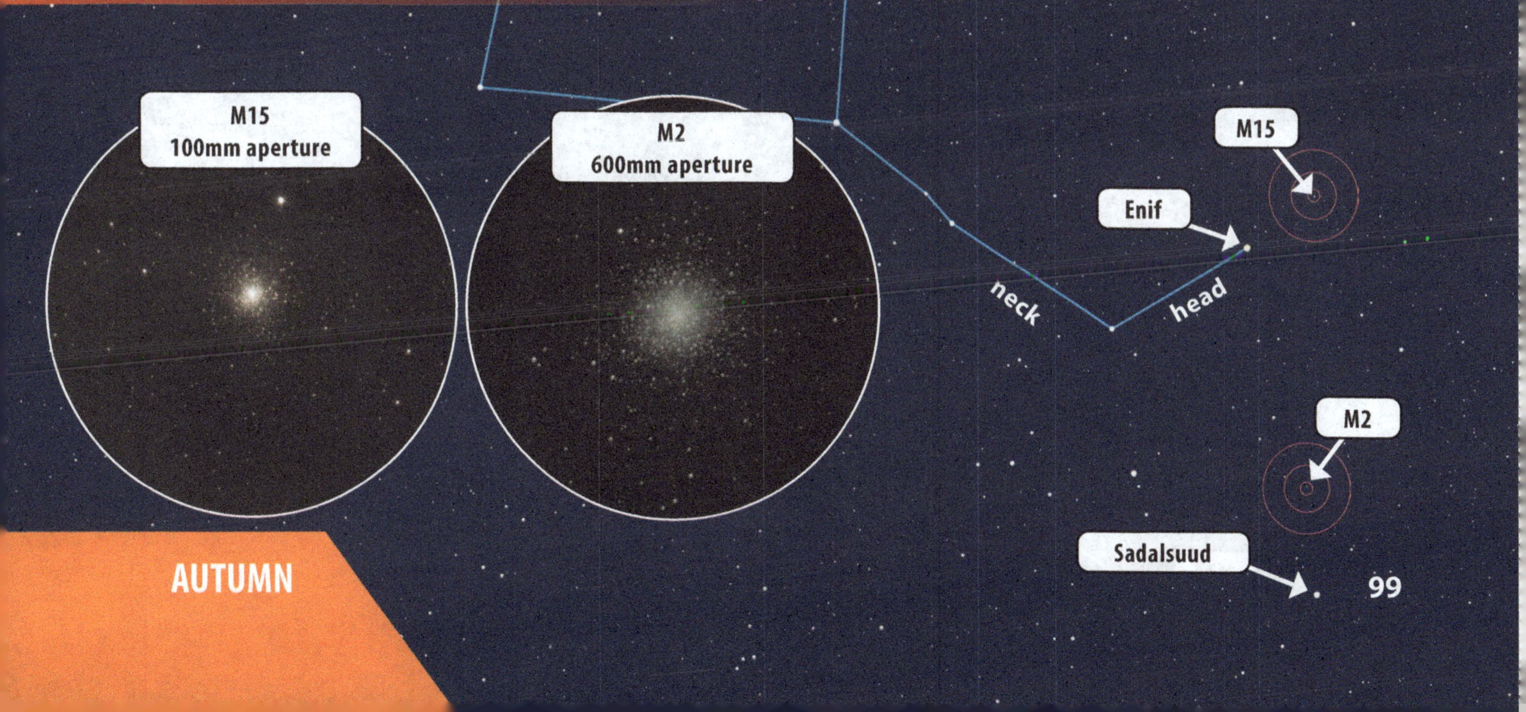

AUTUMN

Spiral Cluster & Alpha Persei Cluster (M34 & Melotte 20)

Surrounding the star Mirfak is a massive star cluster, known as the Alpha Persei Cluster, or Collinder 39. Through a telescope, I consider this cluster to rival the Pleiades (page 110), although it is a far more impressive target in binoculars.

M34, nicknamed the "Spiral Cluster," is a classic open cluster. Interestingly, M34 is equally as bright as M33, the Triangulum Galaxy. M34's light is concentrated over a few dozen bright stars, whereas the galaxy's light is spread over a large area. In other words, even at the same magnitude, M34 is far easier to see.

M34 should be viewed at the lowest magnification your telescope allows. Note how the stars in the cluster seem to be organized in pairs, and these pairs are organized in arcs—the reason for this cluster's nickname.

Observing Log

Date:	Time:	Location:
Sky Conditions (Seeing/Transparency):		
Instrument (Magnification/Aperture):		
Notes:		

Sketch

Object of Interest: The Spiral Cluster (M34)
Dark Sky Requirement: Suburbs
Object Type: Open cluster
Brightness: 5.2
Distance: 1,600 light years
Apparent Size: 35 arcminutes

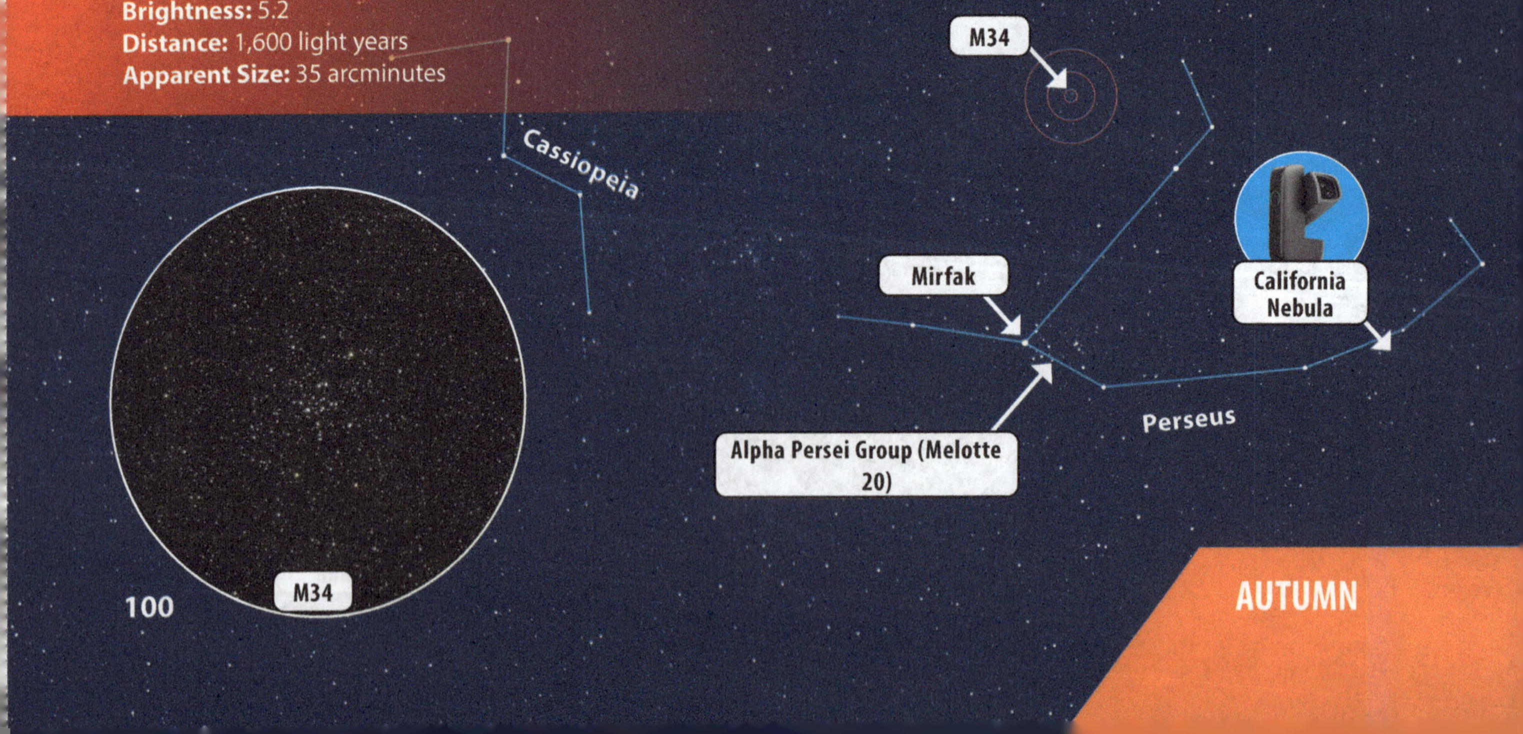

Object of Interest: The Alpha Persei Cluster (Melotte 20)
Dark Sky Requirement: Suburbs
Object Type: Open cluster
Brightness: 2.3
Distance: 600 light years
Apparent Size: 300 arcminutes

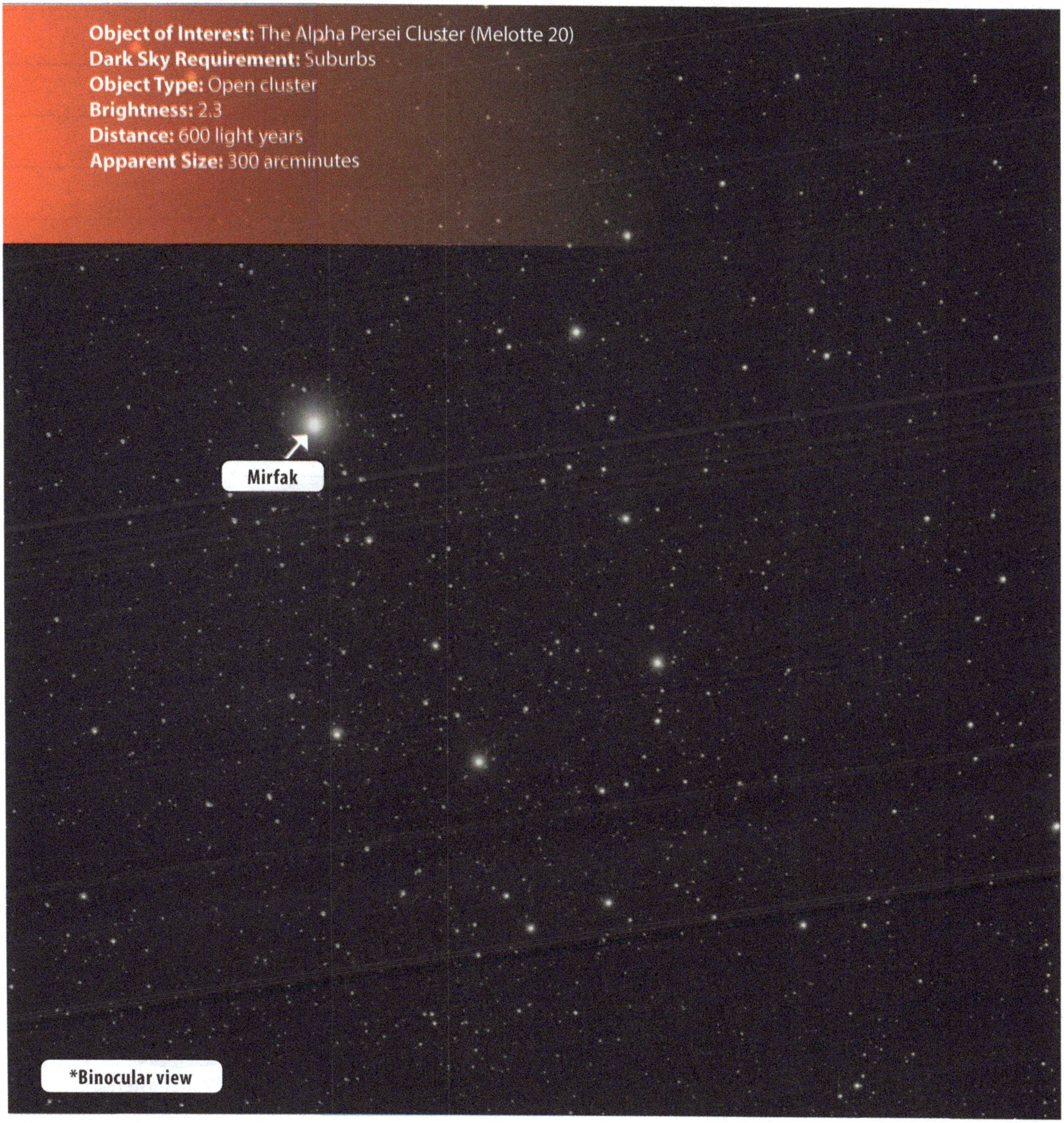

Autumn Double Stars and Asterisms

The autumn sky is host to several lesser known asterisms that can be great to keep in your back pocket to impress folks at your local star party. You might ask a seasoned astronomer, "Did you know about the Question Mark?" and then call them over to your telescope to take a look. It's really quite something to observe someone who's been skywatching for decades as they see something new for the first time.

As for double stars, these are great for adding to your observing log book. As mentioned earlier in this book, astronomy clubs often include double star observing as part of their observing challenge programs, and it's cool to earn certificates (and respect) from your stargazing peers. It's also a chance to observe objects in one of the least visited parts of the sky, Capricornus, where you can find the double stars Alpha Cap and Beta Cap.

There are also several asterisms that require no telescope at all, such as the Circlet and Head of the Whale. However, these are more for helping you navigate around than as specific telescope targets.

Observing Log

Date:	Time:	Location:

Sky Conditions (Seeing/Transparency):

Instrument (Magnification/Aperture):

Notes:

Sketch

Home Plate asterism

Question Mark asterism

Golf Putter asterism

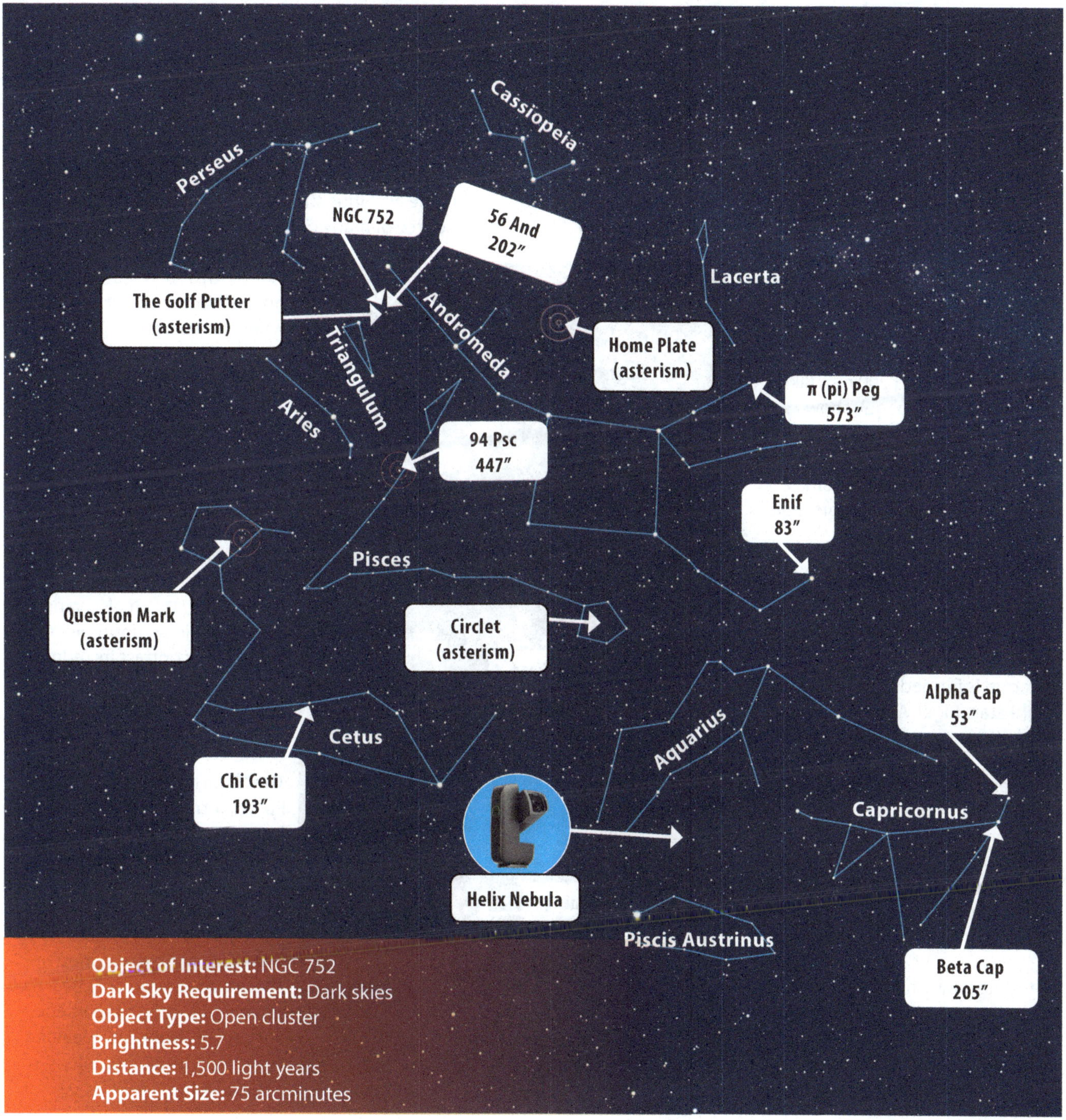

Object of Interest: NGC 752
Dark Sky Requirement: Dark skies
Object Type: Open cluster
Brightness: 5.7
Distance: 1,500 light years
Apparent Size: 75 arcminutes

Winter Targets

Winter stargazing can be particularly challenging, with increased cloud cover and cold temperatures making stargazing unpleasant at best and impossible at worst. That said, ambitious stargazers are rewarded with some of the night sky's most impressive sights, particularly M42—the Orion Nebula—located below Orion's belt, arguably winter's most popular asterism.

Long-exposure photographs, at a focal length around 200 mm (like the image on this page), show considerable structure in this part of the sky. These include several features that cannot be seen with a small telescope, including the Horsehead Nebula.

A large asterism called "the Winter Hexagon" covers almost the entire winter sky, bisecting each of the prominent winter constellations. The Hexagon is formed by joining the bright stars Rigel, Aldebaran, Capella, Pollux, Procyon, and Sirius.

The plane of the Milky Way passes through the middle of the Winter Hexagon. However, this part of the Milky Way faces away from the galactic core. Although the area is speckled with galactic (open) star clusters, it's noticeably less dense.

The winter sky is home to Sirius, the brightest star in the sky. Nicknamed the "Dog Star" due to its prominence in the constellation Canis Major (Greater Dog), Sirius appears in the east just before sunrise in August, inspiring the phrase, "dog days of summer."

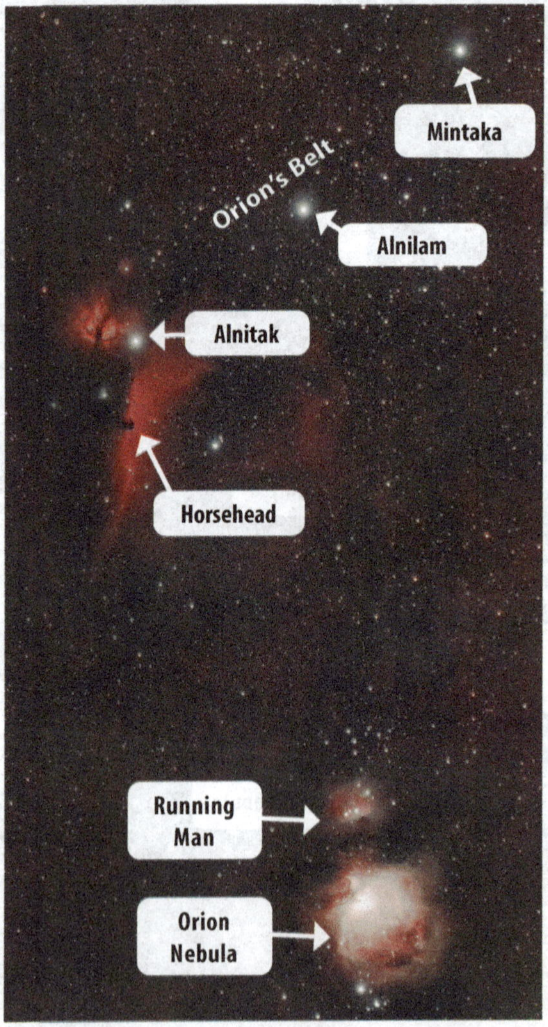

Long exposure photograph of the Orion Cloud Complex

In unstable air, Sirius shimmers in a rainbow of color, and is the source of many of the world's UFO sightings. In a car or airplane, behind light clouds, this star creates the optical illusion that it follows you wherever you go.

The next most prominent winter star is Betelgeuse, somewhere in the vicinity of which *The Hitchhiker's Guide to the Galaxy* is said to have been written. Betelgeuse can be found near the top of the constellation Orion. When looking at this star through a telescope, it's easy to see how red it is. It also varies in brightness in an unpredictable pattern. To contrast its redness, pan the telescope down to Rigel, a blue star located below Orion's Belt.

The "Twins," stars Castor and Pollux, form the head of Gemini. Castor is a double star when viewed through a telescope but a sextuple star system—six stars bound together by gravity, according to data from professional observatories. These six stars can only be separated through the science of **spectroscopy** (i.e., breaking down light into different wavelengths).

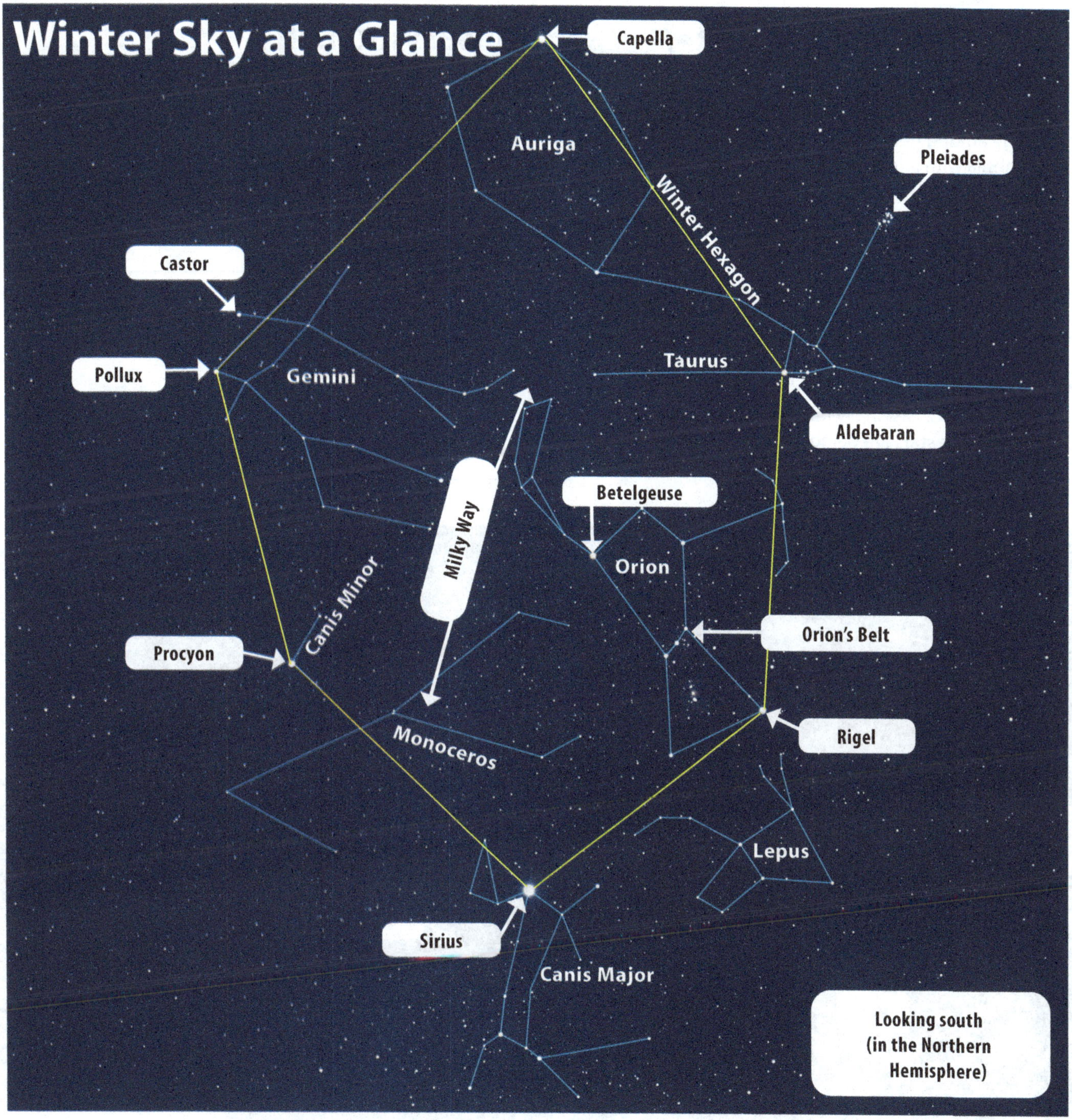
Winter Sky at a Glance
Capella
Auriga
Pleiades
Winter Hexagon
Castor
Pollux
Gemini
Taurus
Aldebaran
Milky Way
Betelgeuse
Orion
Canis Minor
Procyon
Orion's Belt
Rigel
Monoceros
Lepus
Sirius
Canis Major
Looking south (in the Northern Hemisphere)

The Orion Nebula (M42)

The Orion Nebula, M42, is so bright, it can be seen even without a telescope in suburban skies as a small smudge among neighboring stars. When you observe this nebula with a telescope, you'll see a great expanse of gas surrounding tight groupings of stars. At higher magnifications, four bright stars form a pattern known as the Trapezium Cluster.

The Orion Nebula is one of the easiest to find deep-sky objects. To locate M42, find Orion's belt, and then imagine his sword as the line of stars running down from the belt. The middle of this sword contains the Orion Nebula.

This nebula is occasionally referred to as a "star factory" because these stars are being formed out of this mass of gas and dust. Another tiny patch of nebula located near M42 is known as "the Running Man Nebula."

Observing Log

Date:	Time:	Location:

Sky Conditions (Seeing/Transparency):

Instrument (Magnification/Aperture):

Notes:

Sketch

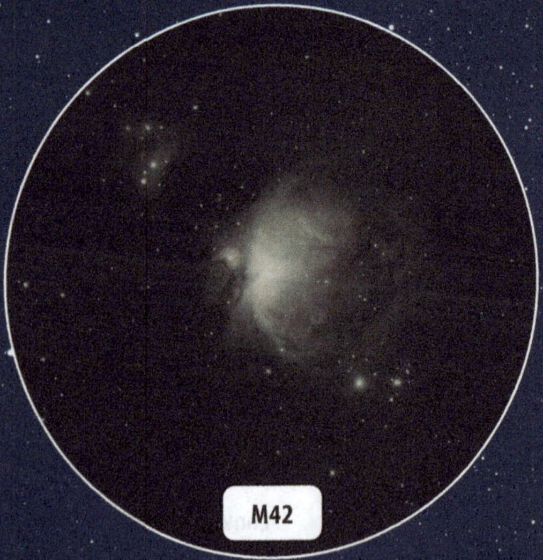

Object of Interest: The Orion Nebula (M42)
Dark Sky Requirement: City-viewable
Object Type: Star-forming nebula
Brightness: 4.0
Distance: 1,400 light years
Apparent Size: 85 × 60 arcminutes

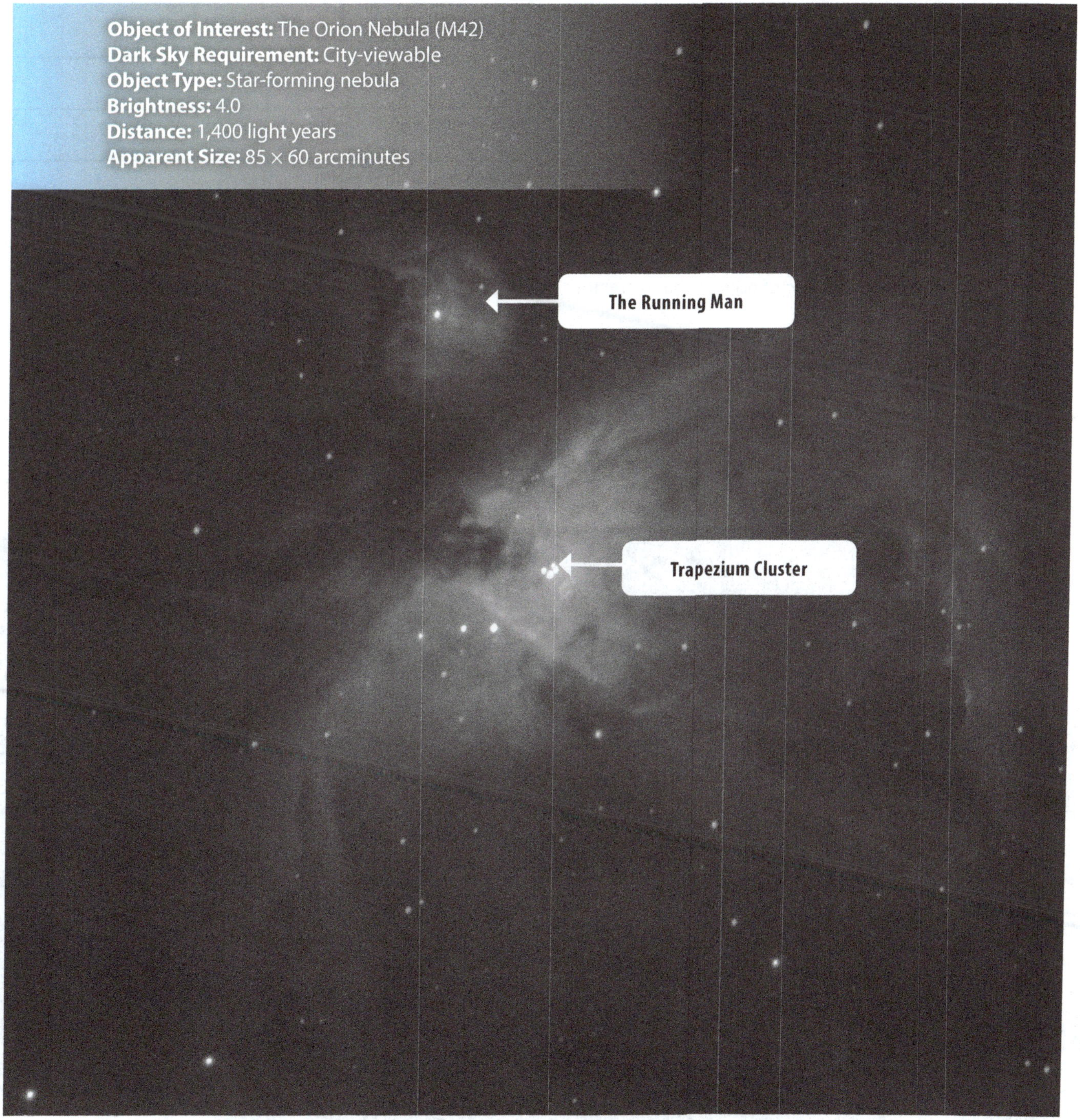

The Crab Nebula (M1)

Something special happened on July 4, 1054. Chinese astronomers recorded what they thought was a new star—a star brighter than Venus! After a few weeks, however, the new star dimmed, but was still visible for almost two years, at which point it was nearly lost to history. But in 1731, the British astronomer John Bevis observed a blob in that exact spot. Almost three decades later, the French comet hunter Charles Messier made this "blob" the first entry in his catalog of objects that were not comets.

The Crab Nebula is a supernova remnant. The Chinese observed the actual supernova—an exploding star. When you look through your telescope at the Crab Nebula, you are observing an ongoing explosion of gas and dust shooting through space at almost five million kilometres per hour.

Observing Log

Date:	Time:	Location:
Sky Conditions (Seeing/Transparency):		
Instrument (Magnification/Aperture):		
Notes:		

Sketch

Object of Interest: The Crab Nebula (M1)
Dark Sky Requirement: Dark skies recommended
Object Type: Supernova remnant
Brightness: 8.4
Distance: 6,200 light years
Apparent Size: ~5 arcminutes

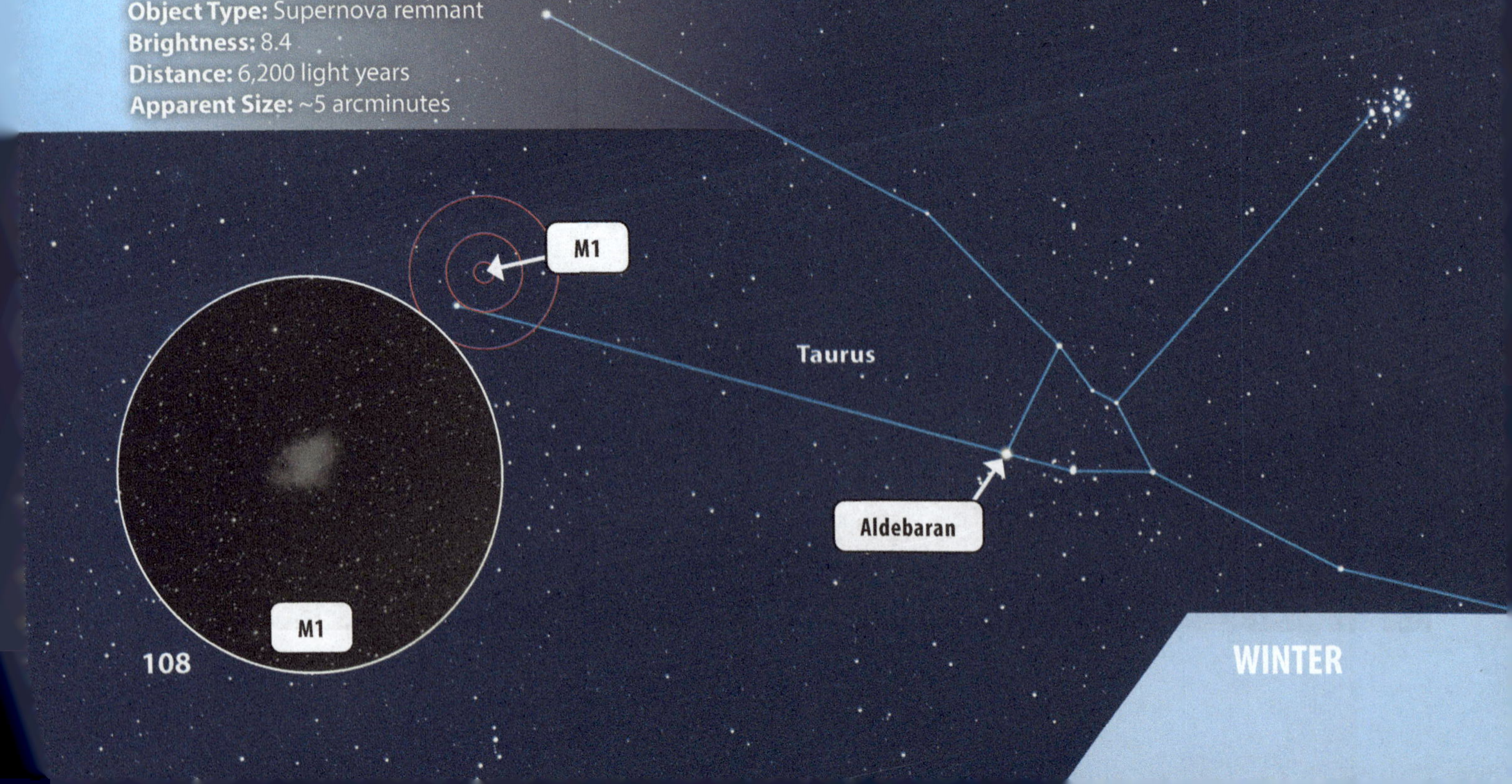

The Shoe Buckle Cluster (M35)

M35 is one of the more popular open clusters in the winter sky. In dark skies, it is visible without a telescope. However, I've sometimes had trouble picking it out in urban skies, even with a telescope.

Start by tracing out the entire constellation of Gemini (the Twins) with your eyes. Focusing on Castor's foot will reveal several bright stars. This is a cascade of stars called "Collinder 89." The Shoe Buckle Cluster is found at the end of the cascade.

If your skies are dark enough, a neighboring star cluster named NGC 2158 will also occupy the same field of view. This is a tight conglomeration of stars, just five arcminutes wide. It almost looks like a distant globular cluster, but is an open cluster, like M35, just about four times as distant.

Observing Log

Date:	Time:	Location:
Sky Conditions (Seeing/Transparency):		
Instrument (Magnification/Aperture):		
Notes:		

Sketch

Object of Interest: The Shoe Buckle Cluster (M35)
Dark Sky Requirement: Suburbs
Object Type: Open cluster
Brightness: 5.1
Distance: 3,000 light years
Apparent Size: 40 arcminutes

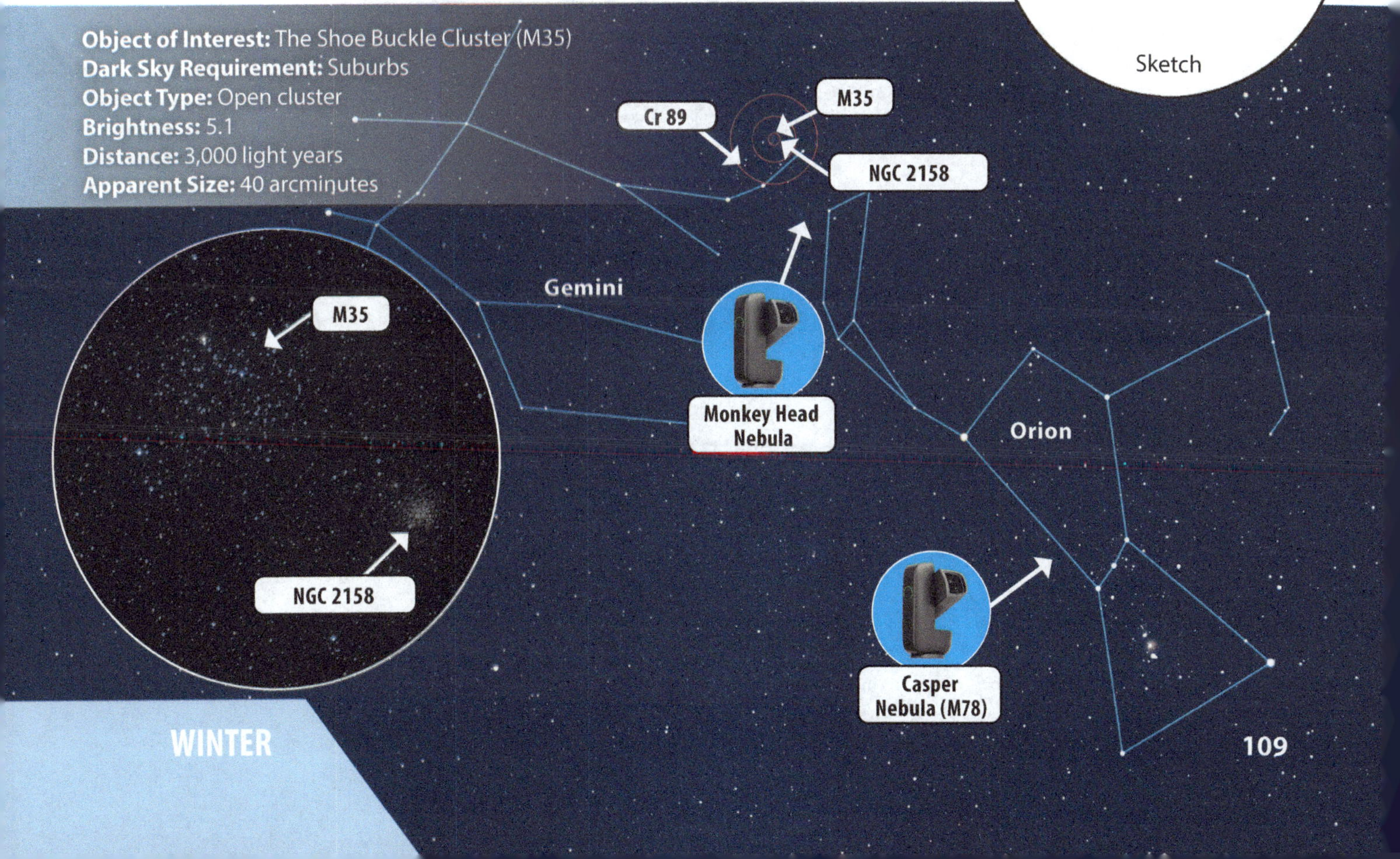

The Pleiades (M45)

The Pleiades can be found to the right of Orion. According to Greek mythology, the Pleiades, or Seven Sisters, were turned into stars by Zeus to help them flee Orion, who still pursues them through the night sky.

Some people think that this cluster of stars is the Little Dipper. It's not. The actual Little Dipper is quite dim, but considerably larger than the Pleiades, and is located in the northern sky.

Usually, with any amount of light pollution, only six of the brightest stars in the Pleiades are visible to the unaided eye. However, as soon as you look in your telescope, dozens more stars will appear! Under extremely dark skies, or in photographs, a patch of reflective dust will appear among the stars.

Observing Log

Date:	Time:	Location:
Sky Conditions (Seeing/Transparency):		
Instrument (Magnification/Aperture):		
Notes:		

Sketch

Object of Interest: The Pleiades (Seven Sisters, Subaru, M45)
Dark Sky Requirement: City-viewable
Object Type: Open cluster
Brightness: 1.5
Distance: 430 light years
Apparent Size: 120 arcminutes

Hyades (Melotte 25)

Residing in the center of the constellation Taurus is the Hyades, the nearest open star cluster to Earth. This makes it a great target to explore with binoculars or a small telescope.

Visually, the cluster is accented by the red giant star Aldebaran. This is simply a chance coincidence. Aldebaran is not actually a part of the cluster at all—at just 67 light years, it lies much closer to Earth.

The cluster also features a bright unaided eye double star, Theta Tauri. On close inspection with a telescope, you should be able to make out a color difference between the two stars of almost equal brightness: one appears slightly orange, while the other is white.

Observing Log

Date:	Time:	Location:
Sky Conditions (Seeing/Transparency):		
Instrument (Magnification/Aperture):		
Notes:		

Sketch

Object of Interest: Hyades (Melotte 25, Caldwell 41)
Dark Sky Requirement: City-viewable
Object Type: Open cluster
Brightness: 0.5
Distance: 147 light years
Apparent Size: 330 arcminutes

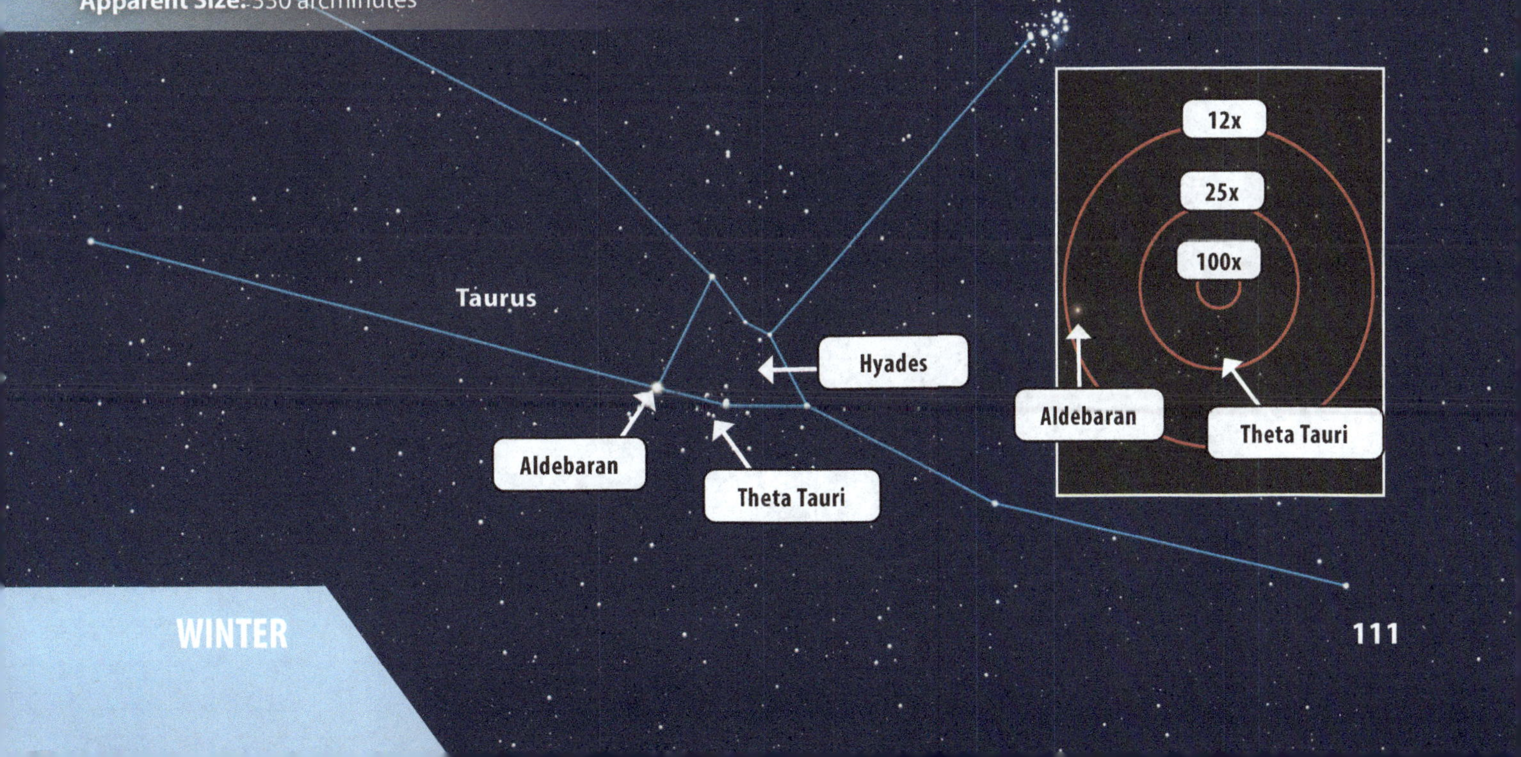

Little Beehive (M41) & M47

M41, the Little Beehive, gets its name from its similarity to M44, the Beehive Cluster, in Cancer. If you can see M41, M44 is most likely also in the sky. Although M44 is categorized as a springtime target, it is quite close to Gemini, a winter constellation. Be sure to compare and contrast these two clusters.

M47 is a tight grouping of several bright (for a cluster) stars wedged between two reddish stars. It's not very dense, but pleasing to the eye.

M46, a larger but dimmer cluster, is found right next to M47 (let's call these the Puppis Clusters, since that's technically where they reside). In extremely dark skies, you may be able to make out a planetary nebula (NGC 2438) within this cluster. There are a lot of other star clusters in the area, so be sure to keep exploring!

Observing Log

Date:	Time:	Location:
Sky Conditions (Seeing/Transparency):		
Instrument (Magnification/Aperture):		
Notes:		

Sketch

Object of Interest: The Little Beehive (M41)
Dark Sky Requirement: Suburbs
Object Type: Open cluster
Brightness: 4.5
Distance: 2,300 light years
Apparent Size: 39 arcminutes

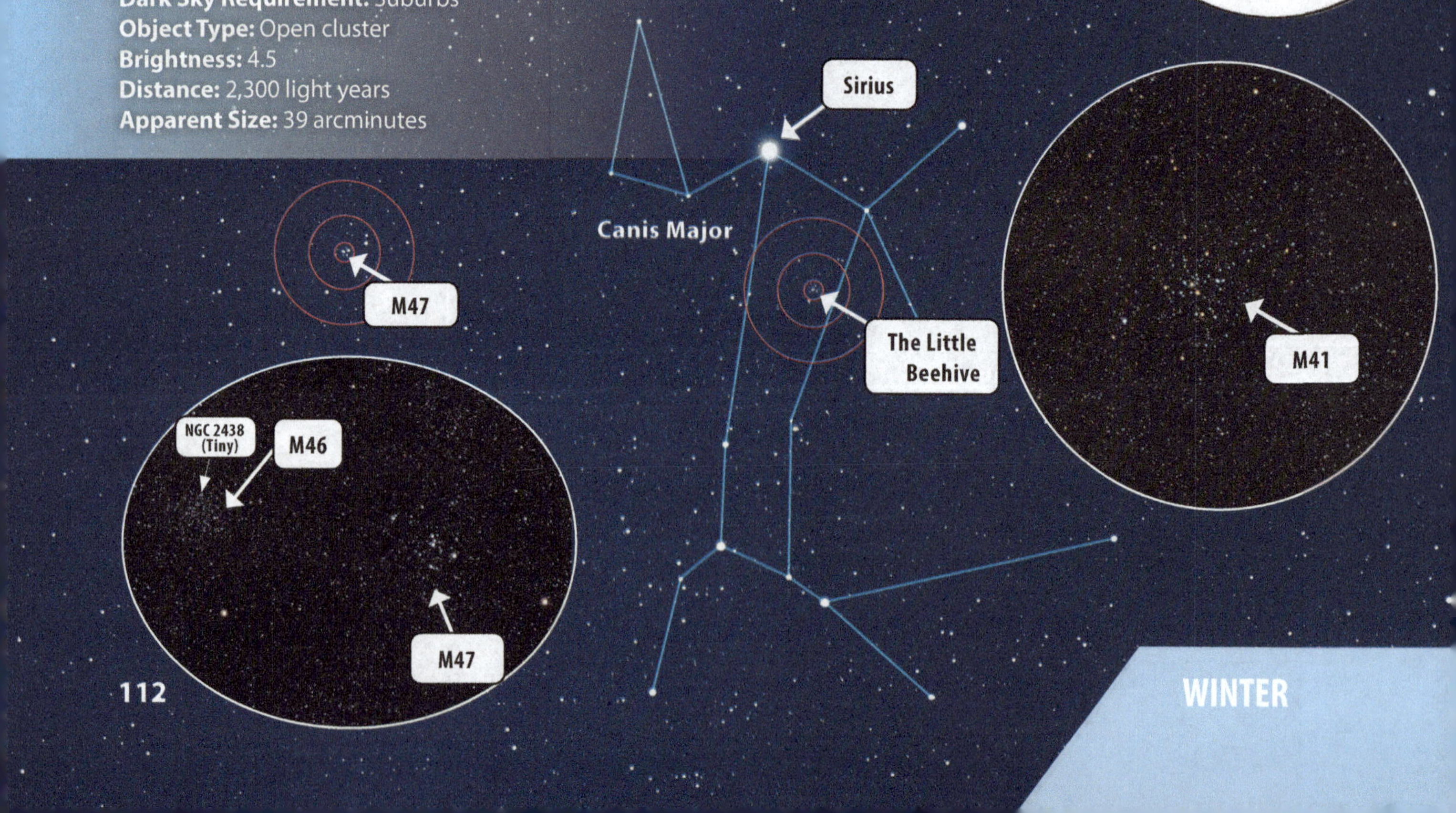

Auriga Clusters (M36, M37, & M38)

The Auriga star clusters can be found by simply pointing your telescope at the middle of Auriga and exploring. Panning back and forth between M37 (the Salt and Pepper Cluster), M36 (the Pinwheel Cluster), and M38 (the Starfish Cluster) allows you to contrast these three clusters nicely.

Auriga is a very busy place, with several colorful cascades of stars and interesting asterisms such as the "Y" and Smiley Face (detailed on page 116).

I've often used these clusters with students at the observatory to come up with new names for them. "The Turtle" seems to be a popular choice for M36. I like to call M37 the "Cloaking Warbird" (again from *Star Trek*). Stellarium calls M38 the Starfish, which I think is pretty fitting.

Observing Log

Date:	Time:	Location:
Sky Conditions (Seeing/Transparency):		
Instrument (Magnification/Aperture):		
Notes:		

Sketch

Object of Interest: The Starfish Cluster (M38)
Dark Sky Requirement: Suburbs
Object Type: Open cluster
Brightness: 6.4
Distance: 4,600 light years
Apparent Size: 20 arcminutes

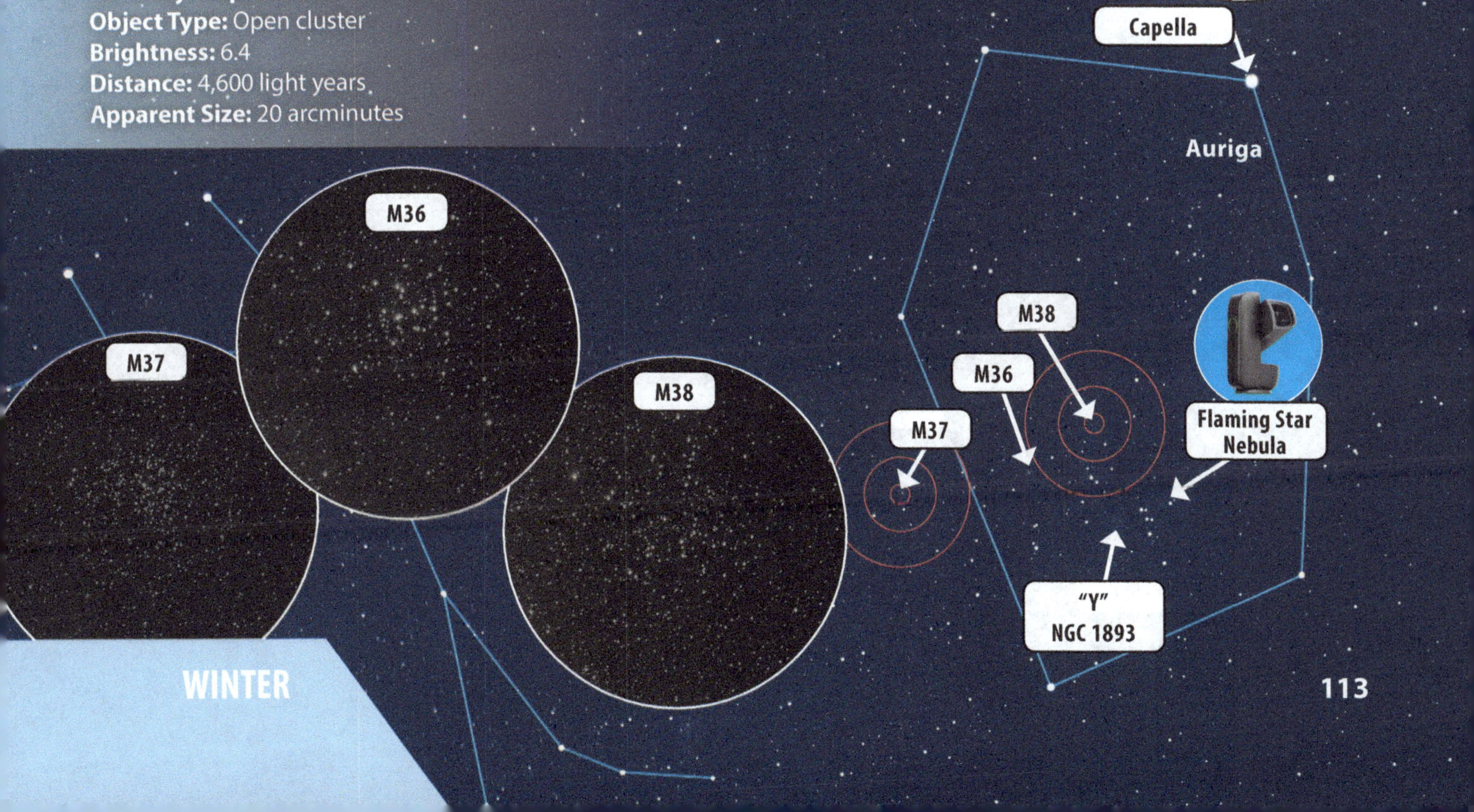

Winter Albireo (145 CMa)

This pair of stars is arguably one of the most beautiful doubles due to their contrasting colors, which some say rival Albireo, the famous summertime double star mentioned earlier in this book.

That said, this star is vastly underrated. During the hundreds of stargazing events I participated in while living in California, I don't recall anyone ever showcasing these particular stars. As of 2025, even the popular stargazing software *SkySafari* only mentions 145 Canis Majoris in the description, and the star map and data entry simply refer to it as "HR 2764." It's not exactly attracting attention to itself!

I find 145 CMa by simply forming a triangle with the hind section of Canis Major.

Observing Log

Date: | **Time:** | **Location:**

Sky Conditions (Seeing/Transparency):

Instrument (Magnification/Aperture):

Notes:

Sketch

Object of Interest: 145 Canis Majoris
Dark Sky Requirement: City-viewable
Object Type: Double star
Brightness: 4.8 and 5.8
Distance: 1400 and 303 light years
Apparent Size: Point source(s)

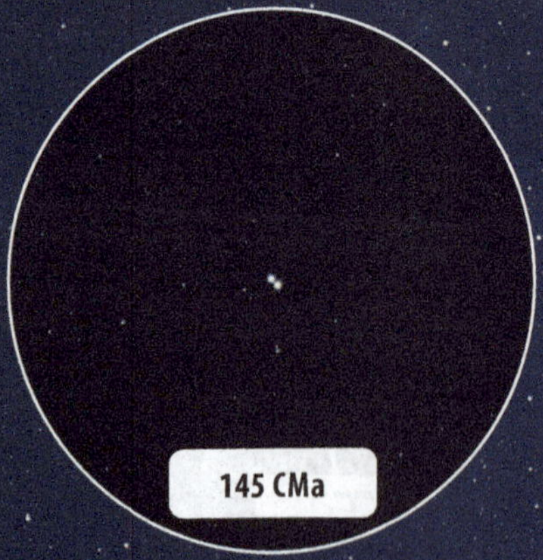

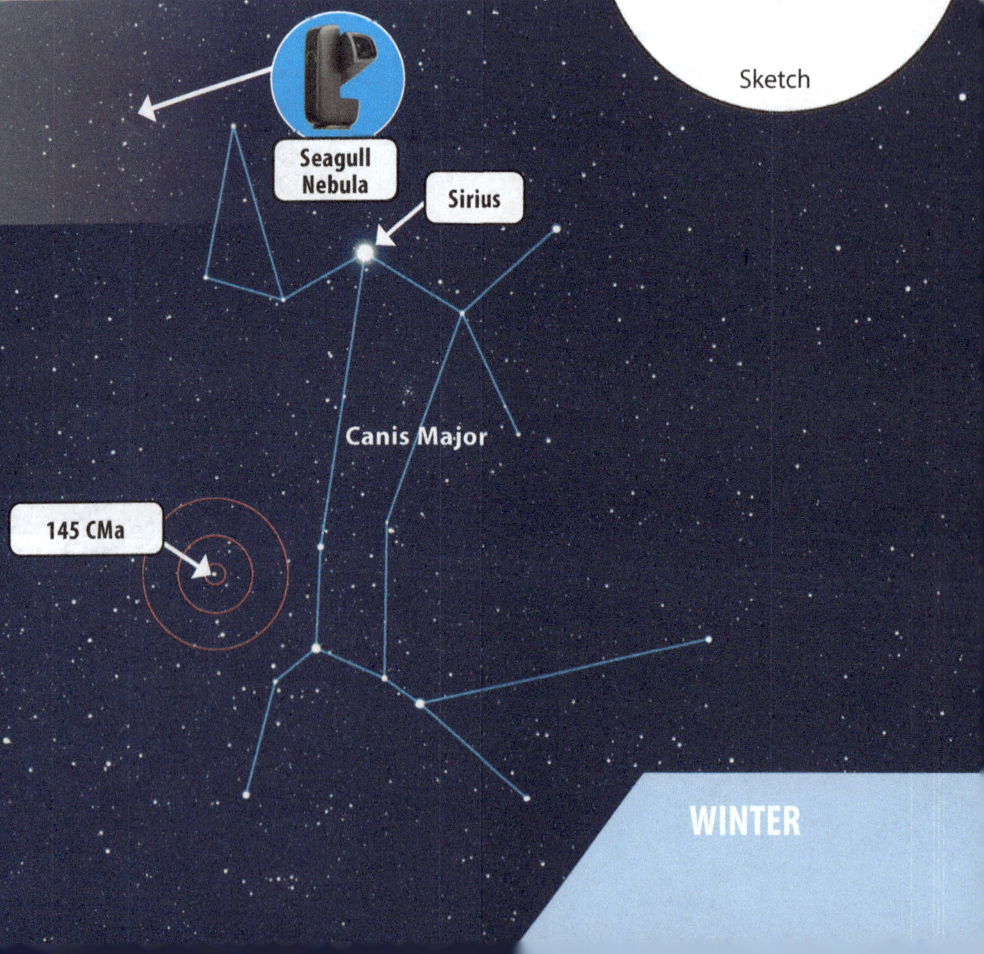

The Christmas Tree Cluster (NGC 2264)

This is a fitting name for a cluster that's high in the sky during the winter. This cluster of stars is shaped like a Christmas tree, with a bright star (15 Mon) representing the base of the trunk. Unless you're using a Newtonian, it's unlikely the cluster will appear right–side-up in the eyepiece.

This cluster also encompasses the Cone Nebula, located where the star or angel would be placed at the top of the Christmas tree. However, it is extremely unlikely you'll be able to make out the nebula without a large telescope and extremely dark skies.

To find NGC 2264, search the area just south of the left foot of Pollux, a star named Alzirr. You can also trace a line eastward from Bellatrix through Betelgeuse in Orion.

Observing Log

Date:	Time:	Location:
Sky Conditions (Seeing/Transparency):		
Instrument (Magnification/Aperture):		
Notes:		

Sketch

Object of Interest: The Christmas Tree Cluster (NGC 2264)
Dark Sky Requirement: Suburbs
Object Type: Cluster with nebulosity
Brightness: 4.1
Distance: 2,200 light years
Apparent Size: 39 arcminutes

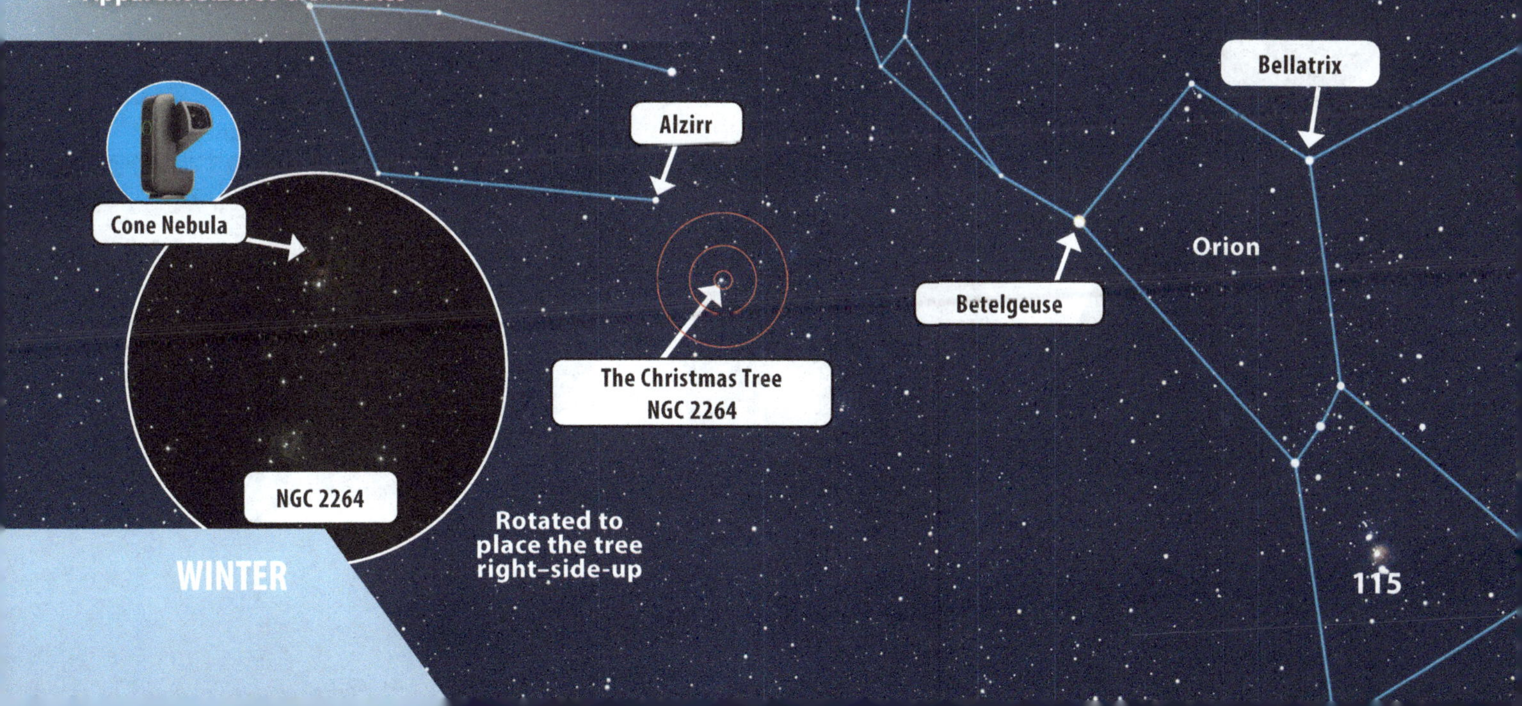

Winter Double Stars and Asterisms

Several of the "double stars" in winter that I return to every so often more closely resemble asterisms. These stars often form very interesting shapes or have different properties, such as significant variations in color. For example, I came across Collinder 69 by chance one evening, calling it "the Robot" in my notes for its similarity in appearance to the "eyes" in the mast of a Mars rover.

Sigma Orionis clearly looks like a sword, and the 37 Cluster, well … you'll just have to see if for yourself. Some are tiny like Lambda-Lambda, and some are huge like Carol's Smiley Face (Carol L. is a frequent poster on the Cloudy Nights online astronomy forum).

Observing Log

Date:	Time:	Location:
Sky Conditions (Seeing/Transparency):		
Instrument (Magnification/Aperture):		
Notes:		

Sketch

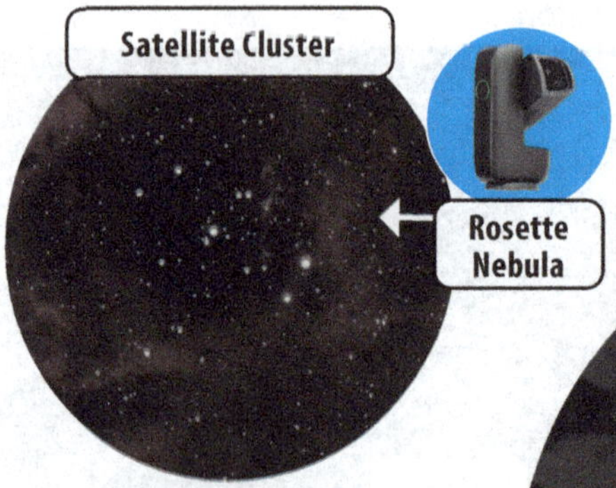

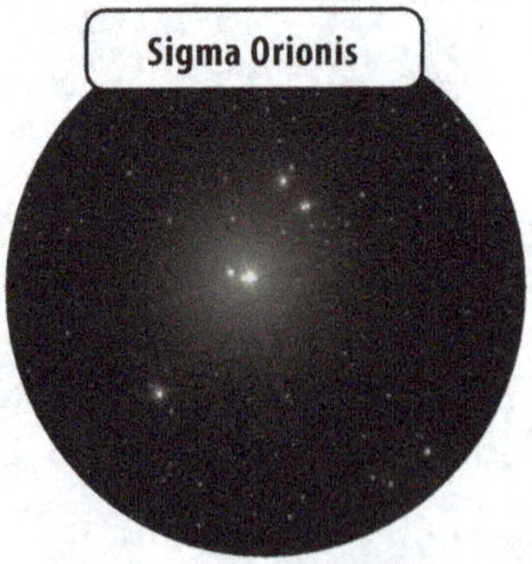

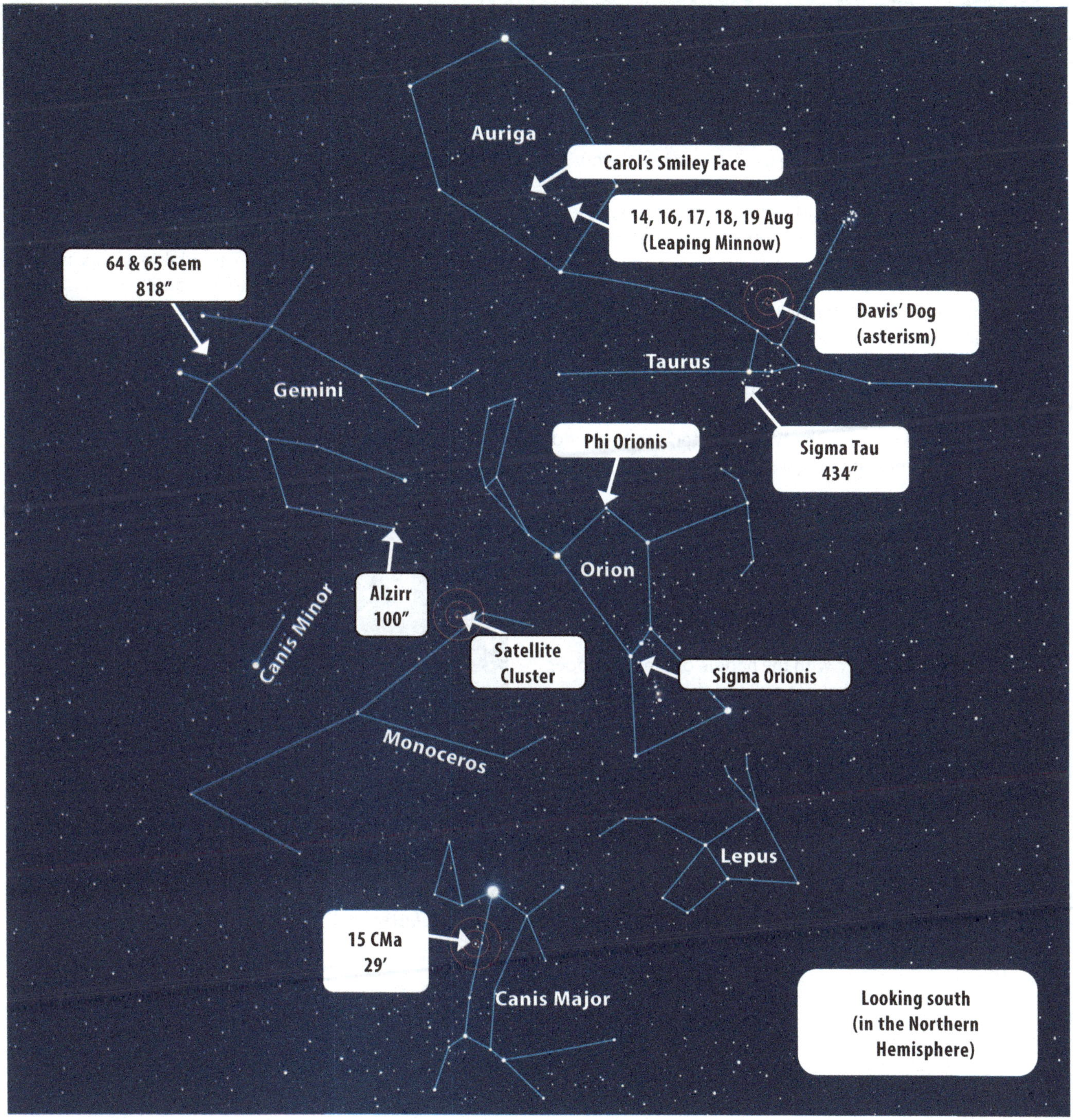
Auriga
Carol's Smiley Face
14, 16, 17, 18, 19 Aug
(Leaping Minnow)
64 & 65 Gem
818″
Davis' Dog
(asterism)
Taurus
Gemini
Sigma Tau
434″
Phi Orionis
Orion
Alzirr
100″
Canis Minor
Satellite
Cluster
Sigma Orionis
Monoceros
Lepus
15 CMa
29′
Canis Major
Looking south
(in the Northern
Hemisphere)

Spring Targets

Springtime is the "short" season for astronomy. The days are getting longer and the nights are getting shorter. It's very noticeable how the lengthening days (and Daylight Saving Time) reduce stargazing opportunities. When observing in springtime, you need either very dark skies or to accept a limited number of targets.

In springtime, the core of the Milky Way doesn't rise until late at night with the summer constellations. This results in a low density of deep-sky objects within our galaxy, such as star clusters and nebulae. What's left are galaxies, which is why astronomers refer to springtime as "Galaxy Season." The largest group of galaxies, known as the "Virgo Cluster," can be found between the constellations Virgo and Leo.

Navigating around the springtime sky is easy thanks to two popular star patterns. The first, if looking south, is the "Spring Triangle." This is formed by joining Regulus, Arcturus, and Spica.

The second pattern is "Arc to Arcturus, Spike to Spica," a phrase that helps to identify two of the three stars in the Spring Triangle. By creating an arc with the handle of the Big Dipper, and following it across the sky to arrive at the bright star Arcturus, you can then straighten your arc to hop over to the bluish star Spica.

Arcturus is an orange giant, and the fourth brightest star in the sky. Spica is a blue giant and the fifteenth brightest star. Spica resides in the constellation Virgo, while Arcturus is located in Boötes.

The most prominent constellations are Leo, Virgo, and Boötes.

Leo contains an asterism known as "the Sickle." Although Leo represents a lion in Greek mythology, it looks like a mouse to me, with the Sickle as the mouse's tail.

Boötes looks like a kite or an ice cream cone with the bright star Arcturus at its base. The brightest stars in Virgo form an asterism known as "the Diamond."

The springtime sky does contain two large galactic (open) star clusters. The first, known as the Coma Cluster, is too large for most telescopes and is better viewed in binoculars. The second cluster, found in the center of the constellation Cancer, is known as the Beehive Cluster.

The Rose Cluster (M5)

Spring Sky at a Glance

Over the course of a human lifetime Arcturus will move relative to nearby stars (about one-seventh the diameter of the Moon).

Spica is both rotating and variable (i.e., it increases and decreases in brightness). At its equator, it rotates at almost 200 kilometres per second, and changes in brightness ever so slightly with each rotation. By comparison, Earth rotates at just under 500 metres per second.

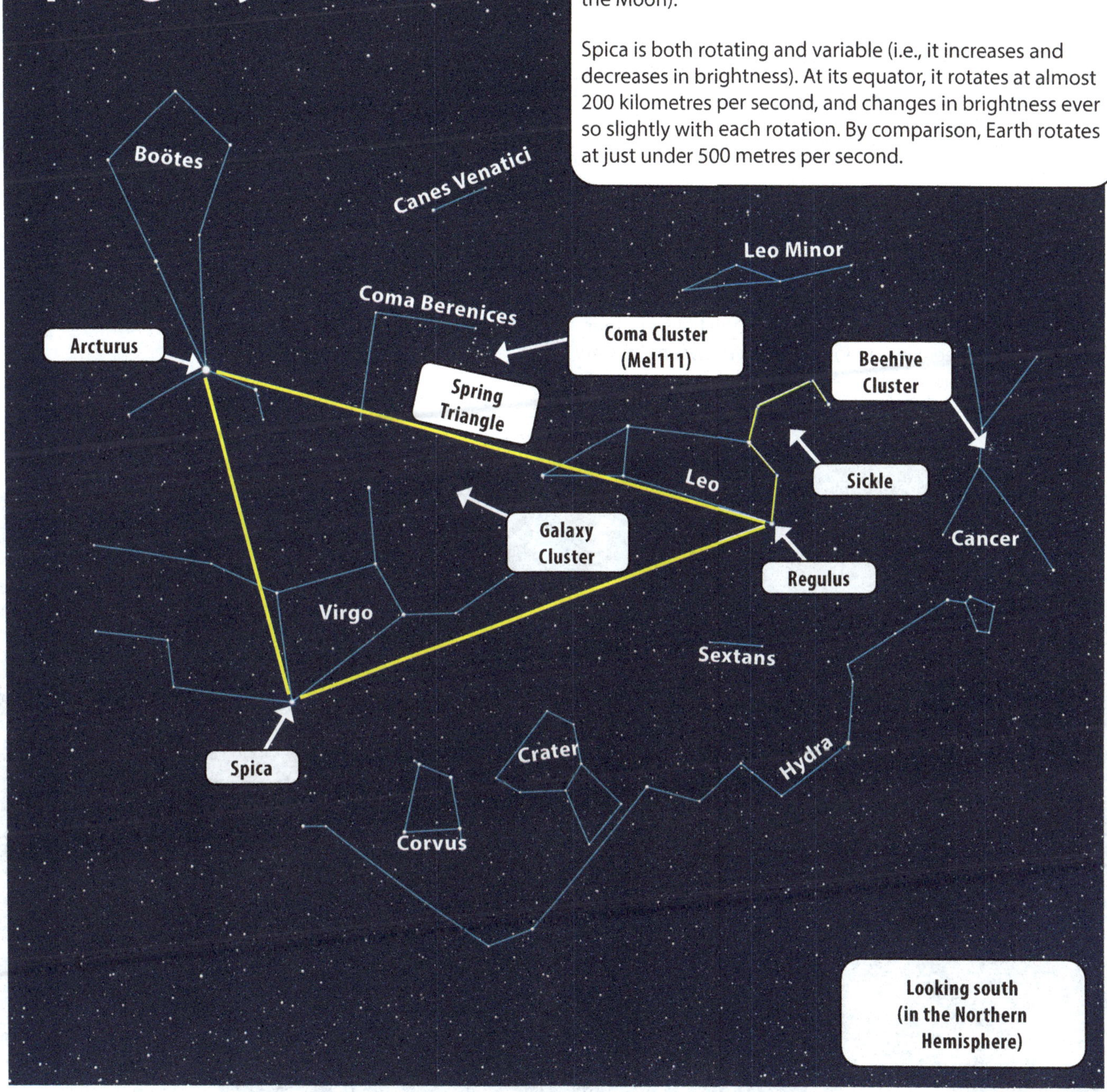

The Beehive Cluster (M44)

The Beehive Cluster is an open cluster that resides in the exact center of Cancer (the Crab). The main stars in Cancer are quite dim, and can be a challenge to identify from the suburbs. The Beehive Cluster itself, however, is brighter than the stars that form the surrounding constellation and can even be seen with the unaided eye from a dark location.

If the stars in Cancer are too dim to see, I usually point my telescope at Leo and simply pan toward the west. The cluster is large enough that you really can't miss it at low magnifications. I identify this cluster by the distinct "house" pattern formed by the brightest five stars. To me, this is meant to represent a bee smoker among the hive. Another way to find the Beehive is to look halfway between the stars Pollux and Regulus.

Observing Log

Date:	Time:	Location:
Sky Conditions (Seeing/Transparency):		
Instrument (Magnification/Aperture):		
Notes:		

Sketch

Object of Interest: The Beehive (M44)
Dark Sky Requirement: City-viewable
Object Type: Open cluster
Brightness: 3.1
Distance: 313 light years
Apparent Size: 120 arcminutes

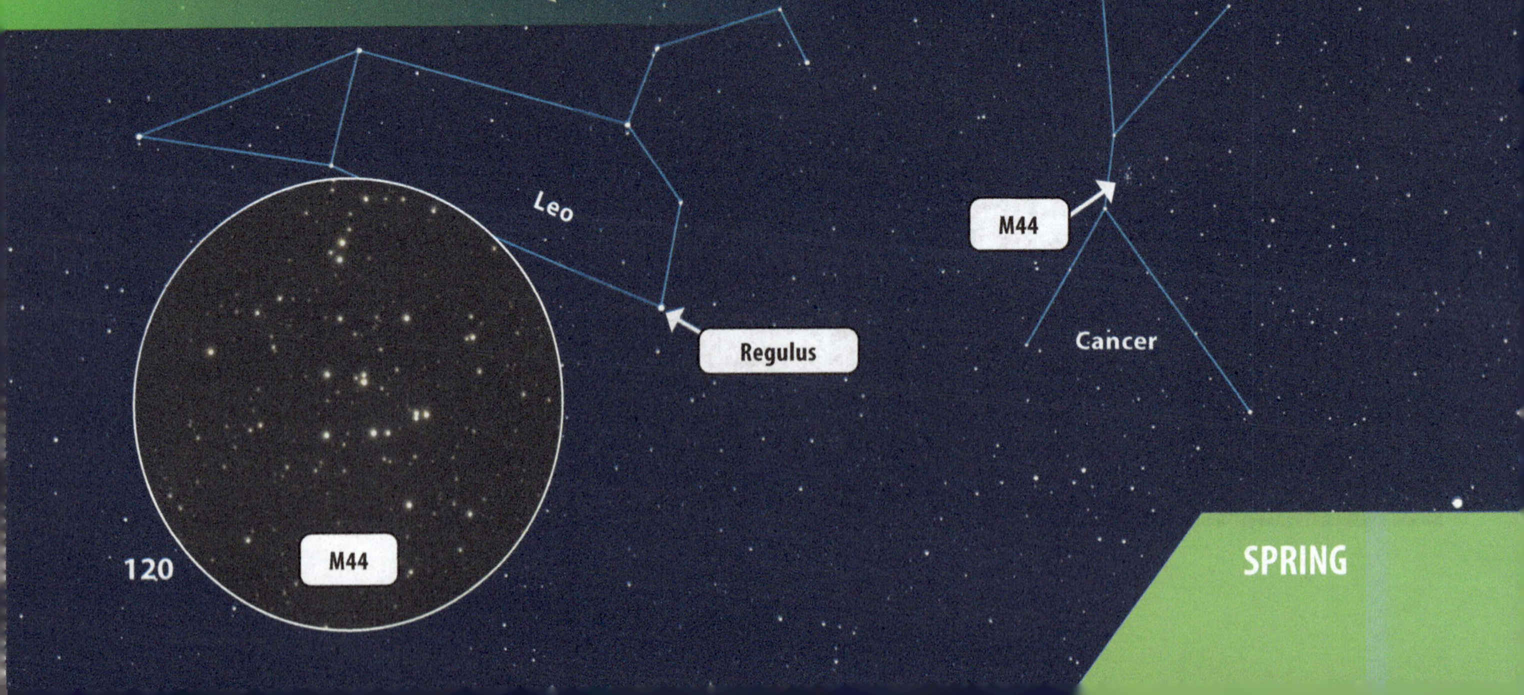

The Sombrero (M104) & Pistol

In spring of 2019, I took my tiny C90 spotting telescope (a telescope no larger than a football) and pointed it at this famous galaxy. I was in a park in Saint John, New Brunswick, and even though we were only a few kilometres from the city center, the galaxy was easy to see.

If you're having trouble finding M104, a nearby star cluster, which cloudynights.com bloggers have dubbed "the Pistol," points straight at the Sombrero Galaxy. You should be able to get the Pistol and M104 in the same field of view.

To find M104, form a triangle with the two bottom stars in the Great Diamond asterism. Alternatively, use your lowest magnification and see if you can identify an asterism called The Arrow. M104 should be (nearly) in the same field of view.

Observing Log

Date:	Time:	Location:
Sky Conditions (Seeing/Transparency):		
Instrument (Magnification/Aperture):		
Notes:		

Sketch

Object of Interest: The Sombrero Galaxy (M104)
Dark Sky Requirement: Suburbs
Object Type: Edge-on spiral galaxy
Brightness: 8.1
Distance: 28 million light years
Apparent Size: 8.4 × 4.9 arcminutes

Globular Cluster (M3)

M3 is a bright globular cluster which is, interestingly, found within the boundaries of the constellation Canes Venatici (whose two brightest stars do not even appear on the map below). I have no trouble identifying this cluster using a small telescope from my backyard in Halifax, Nova Scotia.

To find the cluster, I simply form a triangle between two stars (Arcturus and Rho Boötis) in Boötes.

The cluster lies in the center of a small triangle of relatively bright (ninth magnitude) stars. The cluster itself is huge, containing over 500,000 stars. This is where aperture really matters—the larger the aperture of your telescope, the more stars you will be able to resolve.

Observing Log

Date:

Time:

Location:

Sky Conditions (Seeing/Transparency):

Instrument (Magnification/Aperture):

Notes:

Sketch

Object of Interest: M3
Dark Sky Requirement: Suburbs
Object Type: Globular cluster
Brightness: 6.2
Distance: 33,000 light years
Apparent Size: 18 arcminutes

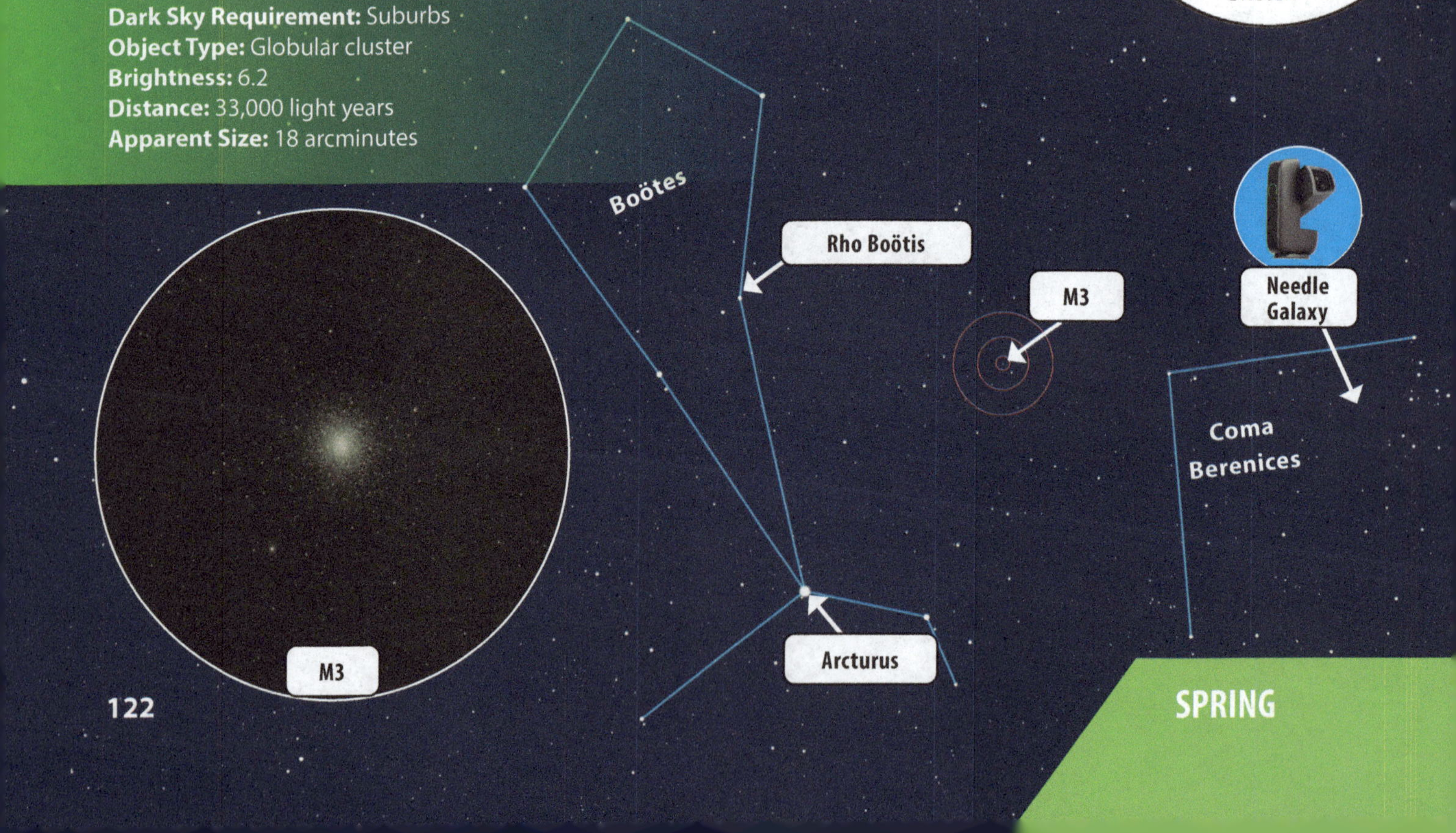

The Rose Cluster (M5)

Globular Cluster M5 is brighter than M3, but arguably slightly more difficult to find. Technically, it's even brighter than M13 (5.7 vs. 5.8, respectively) As mentioned elsewhere, even though this cluster is "bright," it appears dim since it's spread out over a large area.

M5 is found within the same field of view as the relatively bright star 5 Serpentis, which can be seen without a telescope in dark skies. This cluster doesn't actually get its name for its appearance, but possibly from the Earl of "Rosse," an Irish astronomer, from whom this nickname may be derived. These clusters are studied intensively by astronomers. Having a large sample of stars, all of which interact with each other, at a similar distance makes for interesting science in the fields of variable star observations and even novae (stars that borrow/steal matter from other stars and temporarily brighten, appearing as "new" stars).

Observing Log

Date:	Time:	Location:
Sky Conditions (Seeing/Transparency):		
Instrument (Magnification/Aperture):		
Notes:		

Sketch

Object of Interest: The Rose Cluster (M5)
Dark Sky Requirement: Suburbs
Object Type: Globular cluster
Brightness: 5.7
Distance: 24,000 light years
Apparent Size: 23 arcminutes

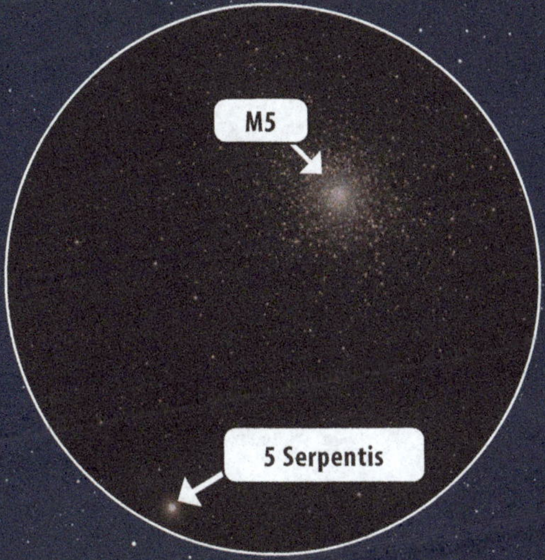

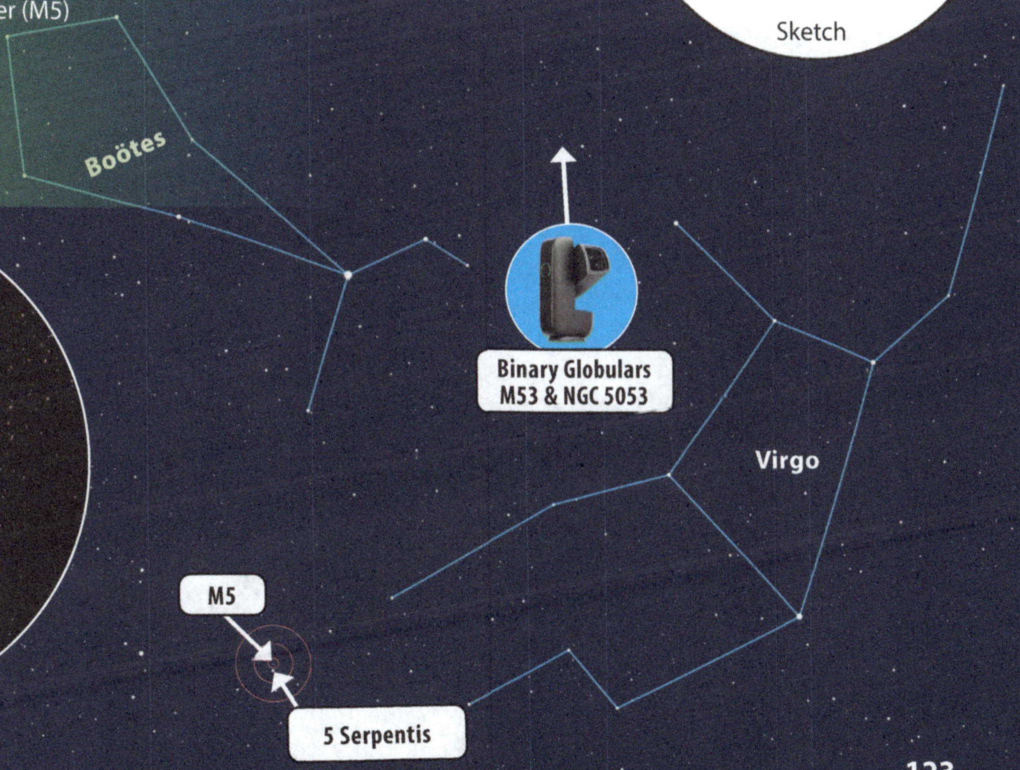

SPRING

The Galaxies of Leo

There are two galaxy trios located just below Leo. The first, known as the Leo Triplet, is more popular. This includes galaxies M65, M66, and NGC 3628 (aka, the Hamburger). While the first two galaxies made Charles Messier's list, the third did not. While the third is slightly less bright than the first two, it's still visible in small telescopes under dark skies. There are two theories for why NGC 3628, and other relatively bright galaxies, did not make Messier's list. The first is that the quality of telescope optics has improved significantly. Second, despite the skies being relatively dark in Paris in the late 1700s, they were often heavy with smoke from chimneys. This chimney smoke may have made the Hamburger Galaxy too difficult to see.

The second trio is simply called the "M96 Group," and consists of Messier objects M95, M96, and M105. Another bright galaxy, missed by Charles Messier is NGC 2903, located near the tip of the Sickle.

Observing Log

Date:	Time:	Location:

Sky Conditions (Seeing/Transparency):

Instrument (Magnification/Aperture):

Notes:

Sketch

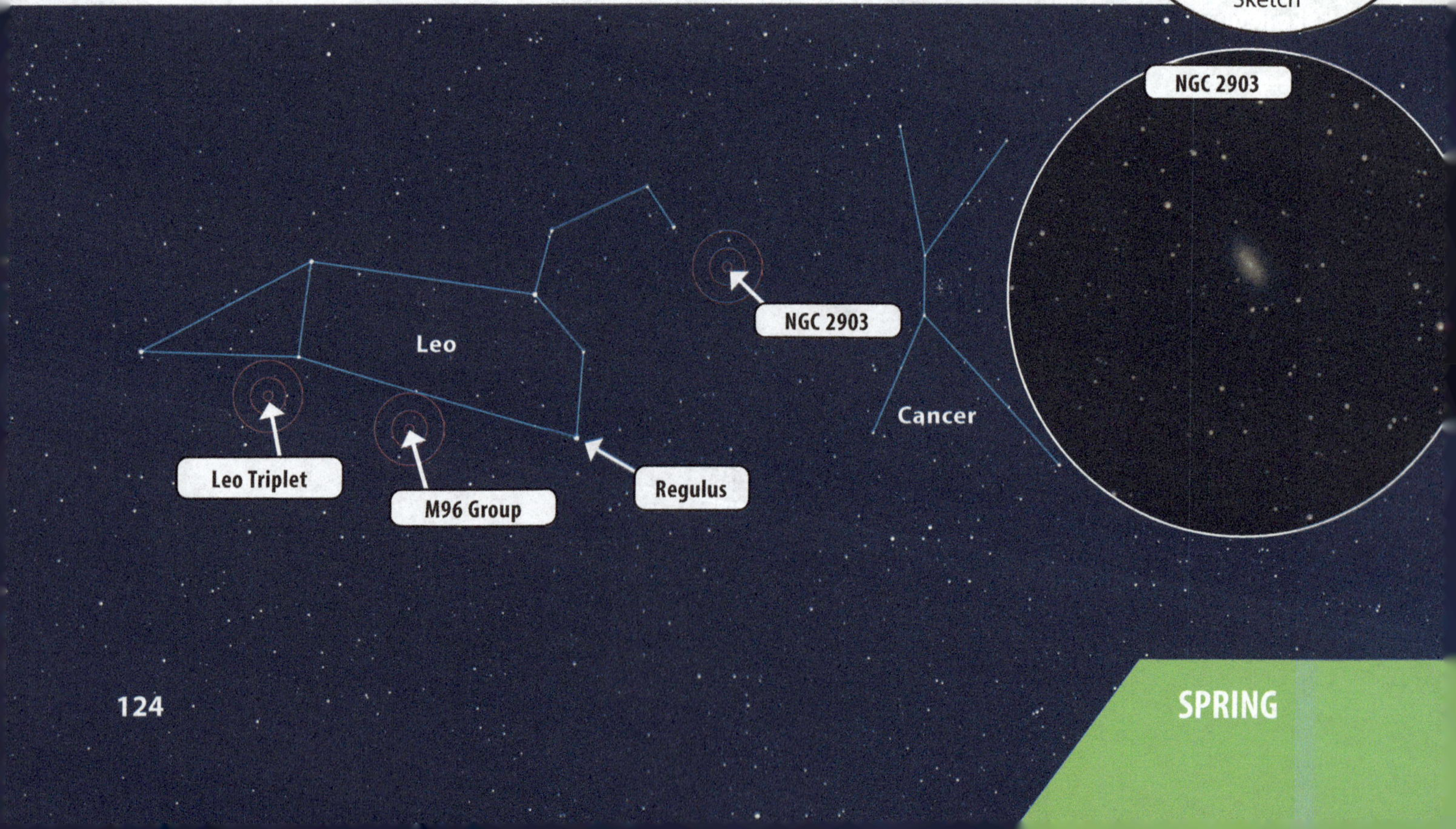

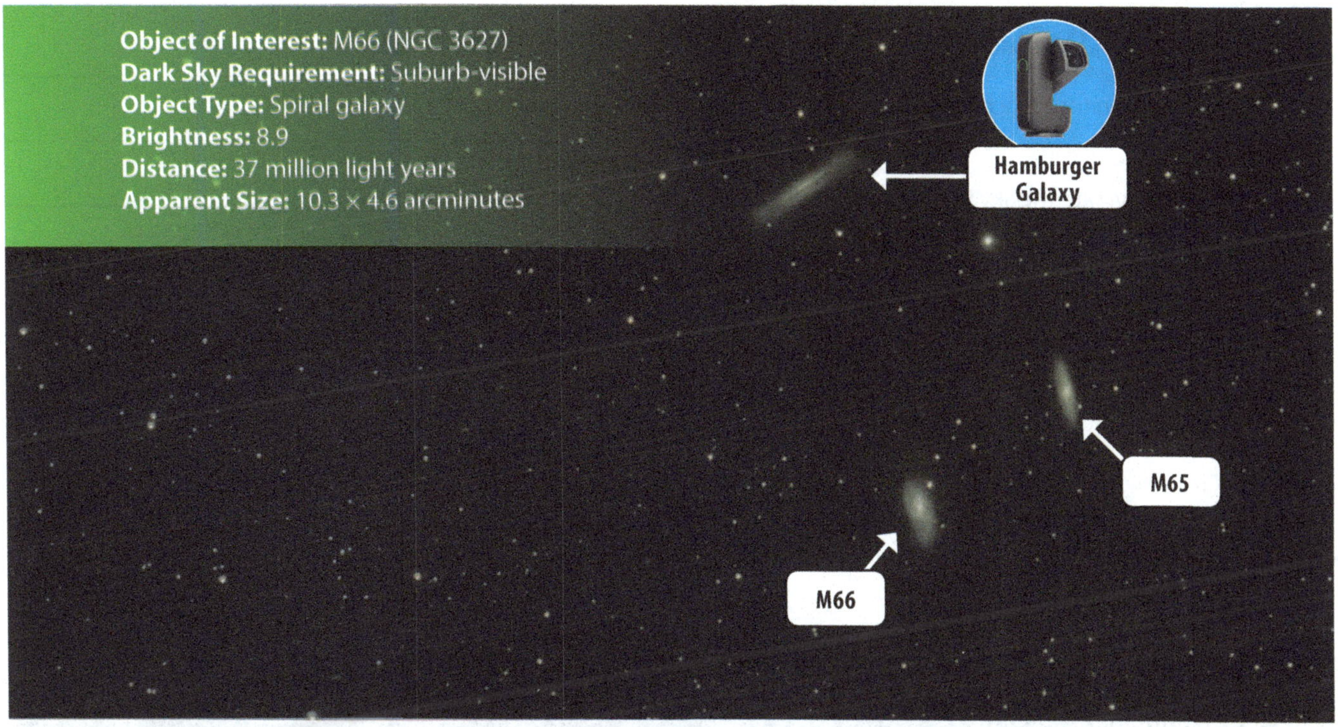
Object of Interest: M66 (NGC 3627)
Dark Sky Requirement: Suburb-visible
Object Type: Spiral galaxy
Brightness: 8.9
Distance: 37 million light years
Apparent Size: 10.3 × 4.6 arcminutes
Hamburger Galaxy
M65
M66

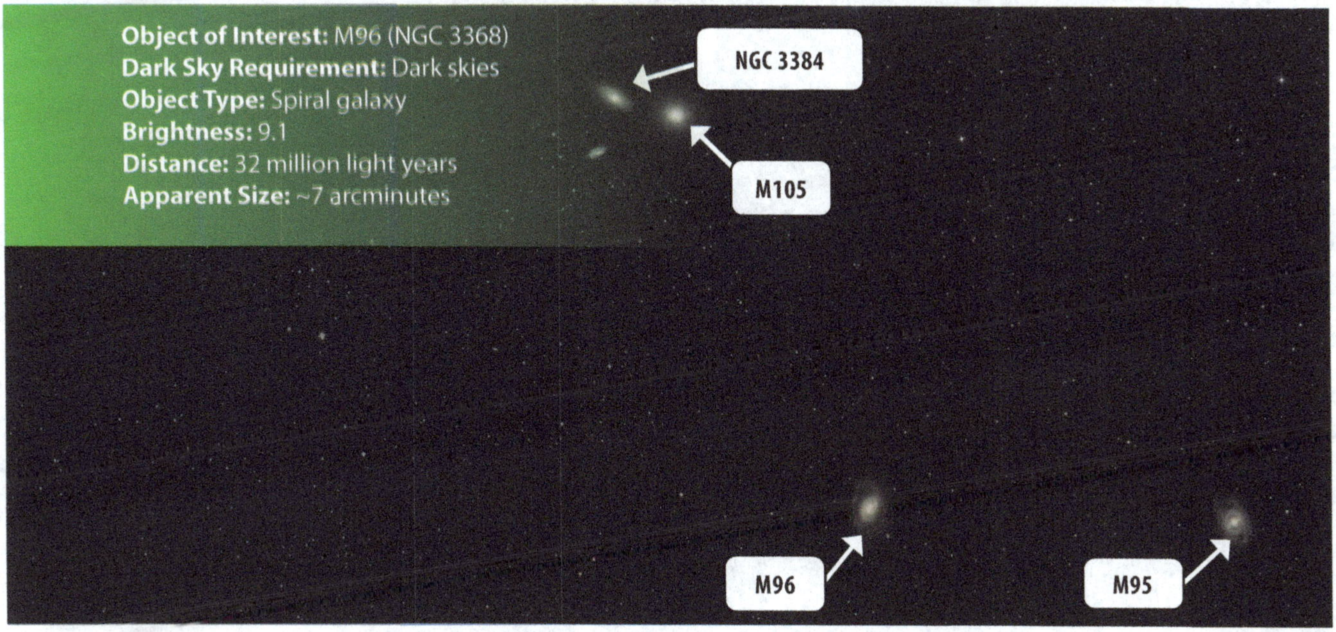
Object of Interest: M96 (NGC 3368)
Dark Sky Requirement: Dark skies
Object Type: Spiral galaxy
Brightness: 9.1
Distance: 32 million light years
Apparent Size: ~7 arcminutes
NGC 3384
M105
M96
M95

The Virgo Cluster

From dark skies, a multitude of galaxies are visible to the amateur astronomer.

Most stargazers navigate the sky through a process they call "star hopping." Within the Virgo Cluster, seasoned astronomers learn how to "galaxy hop." They form a map of the cluster in their minds and are able to move the telescope from one galaxy to another with a light touch on the optical tube. From the suburbs, only the brightest of galaxies are visible, such as M86, which appears as a (relatively) bright smudge against the background sky.

The Virgo Cluster contains Markarian's Chain, a row of about eight galaxies that appear to arc across the sky. The entire cluster contains upward of 2,000 galaxies, each containing billions of stars.

Observing Log

Date:

Time:

Location:

Sky Conditions (Seeing/Transparency):

Instrument (Magnification/Aperture):

Notes:

Sketch

Object of Interest: M87
Dark Sky Requirement: Suburbs
Object Type: Elliptical galaxy
Brightness: 8.7
Distance: 55 million light years
Apparent Size: ~7 arcminutes

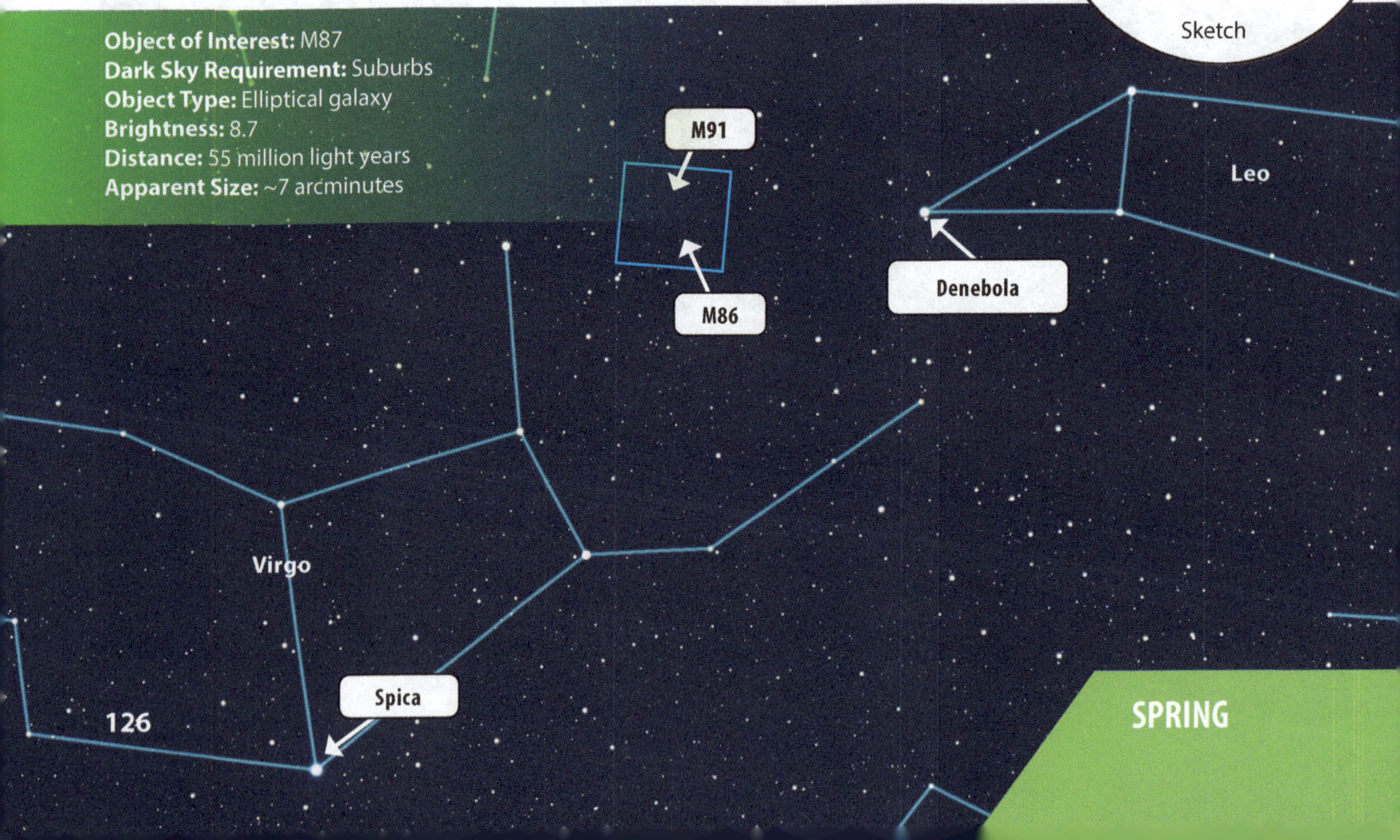

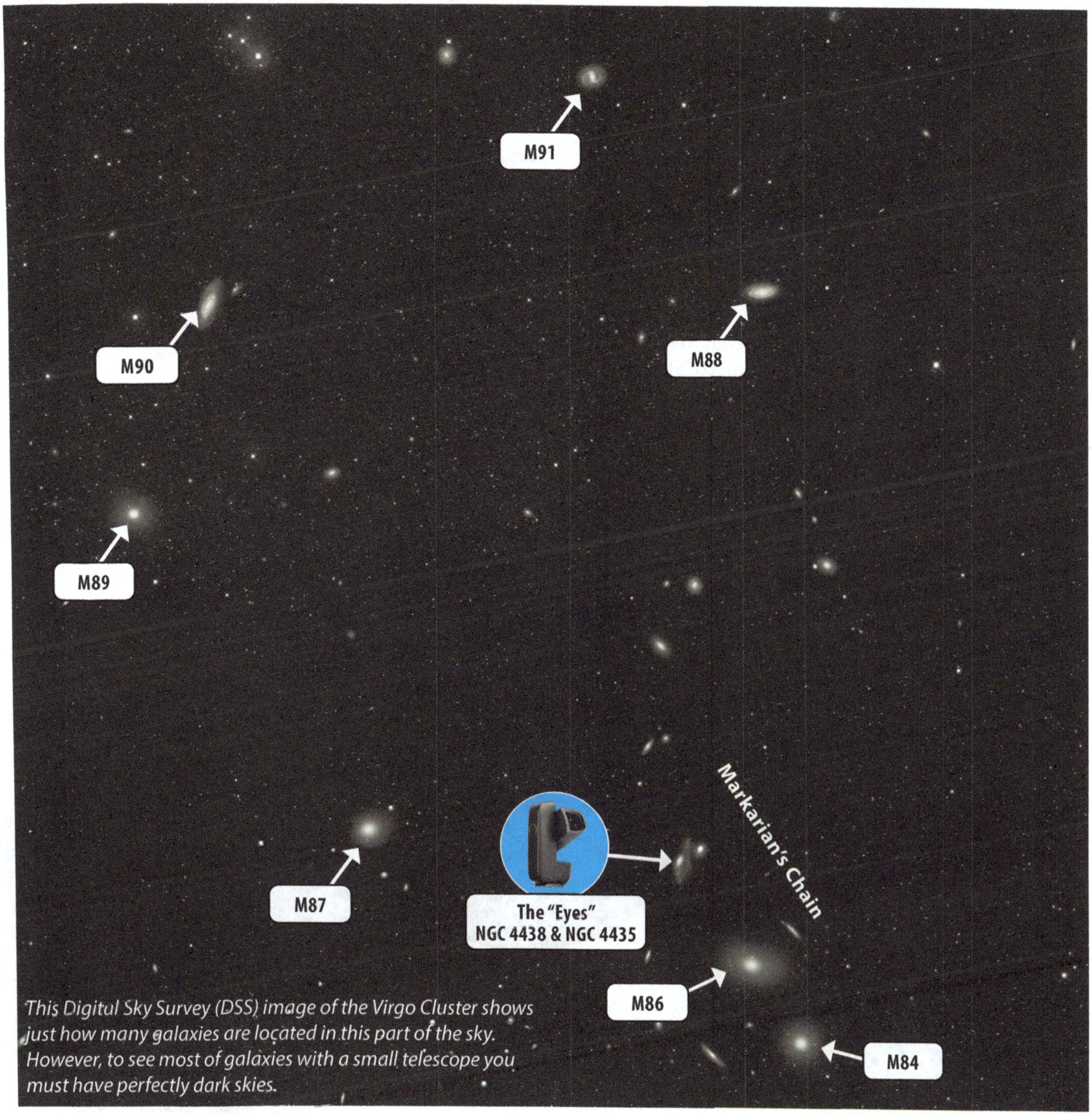

This Digital Sky Survey (DSS) image of the Virgo Cluster shows just how many galaxies are located in this part of the sky. However, to see most of galaxies with a small telescope you must have perfectly dark skies.

Spring Double Stars and Asterisms

While springtime sky is light on star clusters, amateur astronomers have more than made up for this with a collection of interesting springtime asterisms. There is a Seahorse asterism surrounding the star Psi Hydrae; there is the Fly Swatter (with a red star that represents the fly) near Mu Hydrae; and one of my favorites, Napoleon's Hat, which is found right next to Arcturus.

Another interesting asterism I found recently in the astronomical League's observing lists is called "Stargate," and consists of a pair of nested triangles of stars located near the Sombrero Galaxy (M104). There's also "Greg's 3" in Leo, and from an old *Sky & Telescope* article, "the Kangaroo" in the constellation Boötes.

Spring is also a busy season for double stars. RASC's "Explore the Universe" program contains 10 springtime double stars.

Observing Log

Date:	Time:	Location:

Sky Conditions (Seeing/Transparency):

Instrument (Magnification/Aperture):

Notes:

Sketch

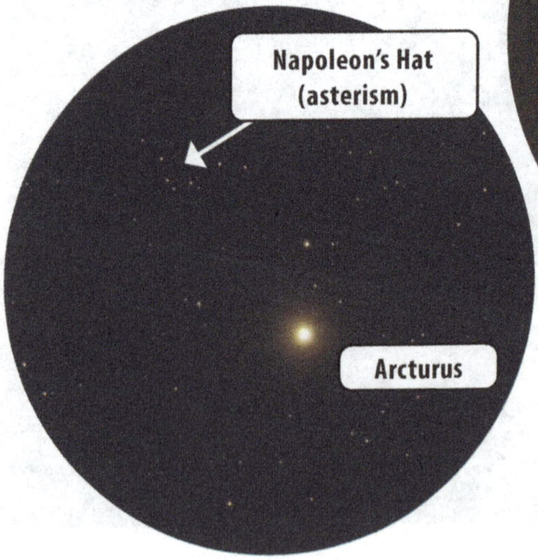

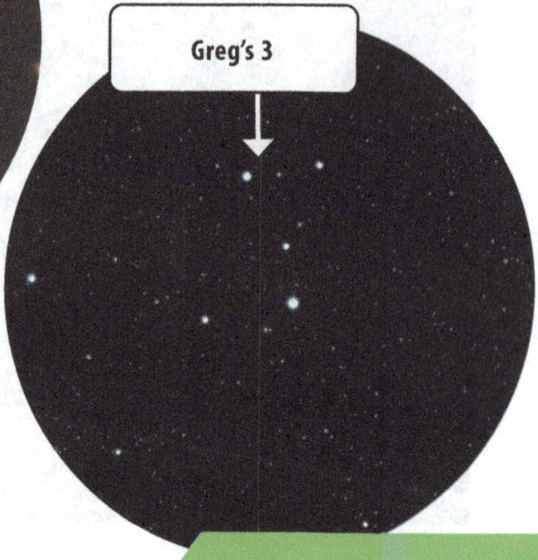

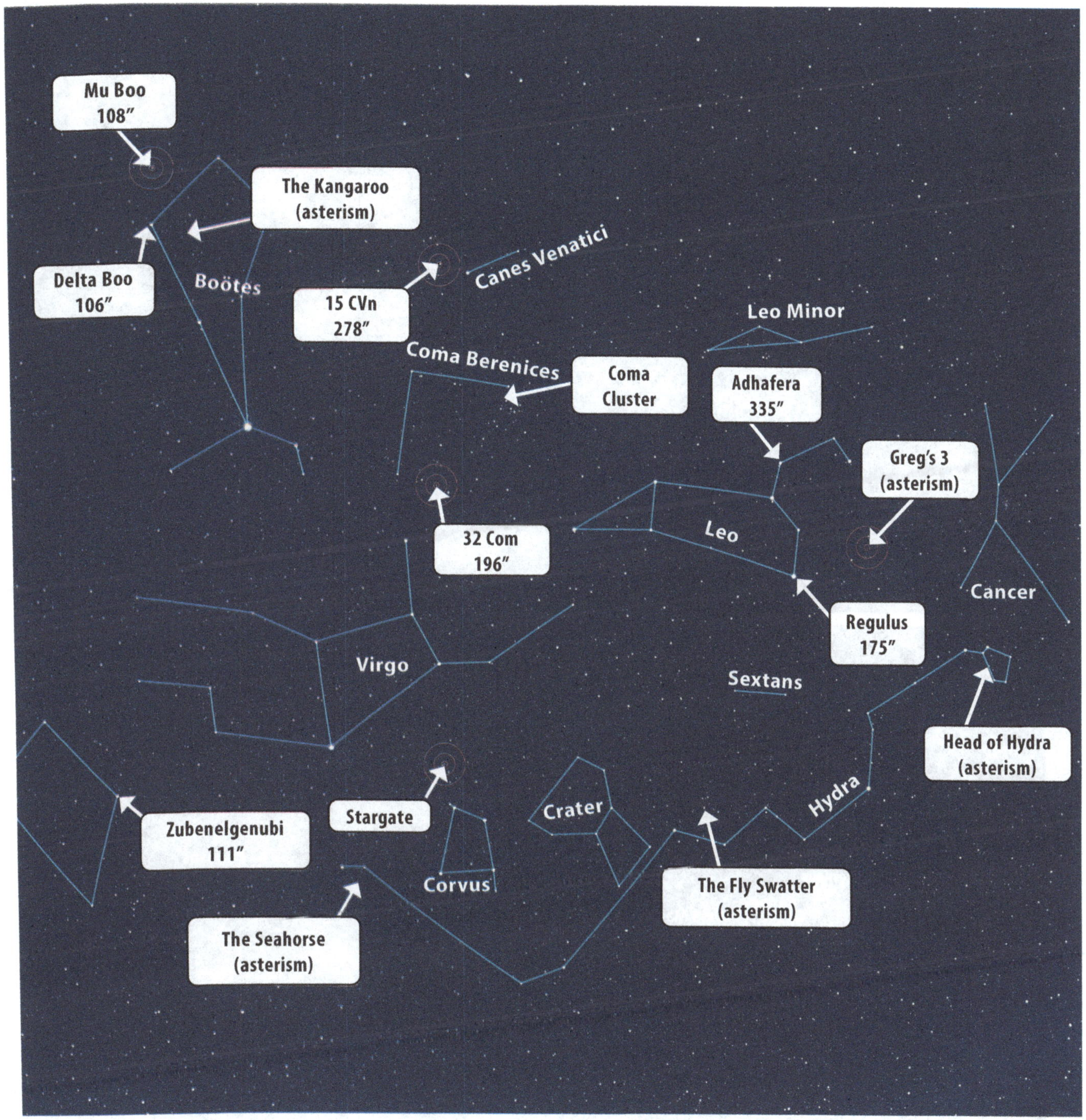
Mu Boo
108″
The Kangaroo
(asterism)
Delta Boo
106″
Boötes
15 CVn
278″
Canes Venatici
Leo Minor
Coma Berenices
Coma
Cluster
Adhafera
335″
Greg's 3
(asterism)
32 Com
196″
Leo
Cancer
Regulus
175″
Virgo
Sextans
Head of Hydra
(asterism)
Hydra
Crater
Stargate
Zubenelgenubi
111″
Corvus
The Fly Swatter
(asterism)
The Seahorse
(asterism)

Summer Full Sky Constellations

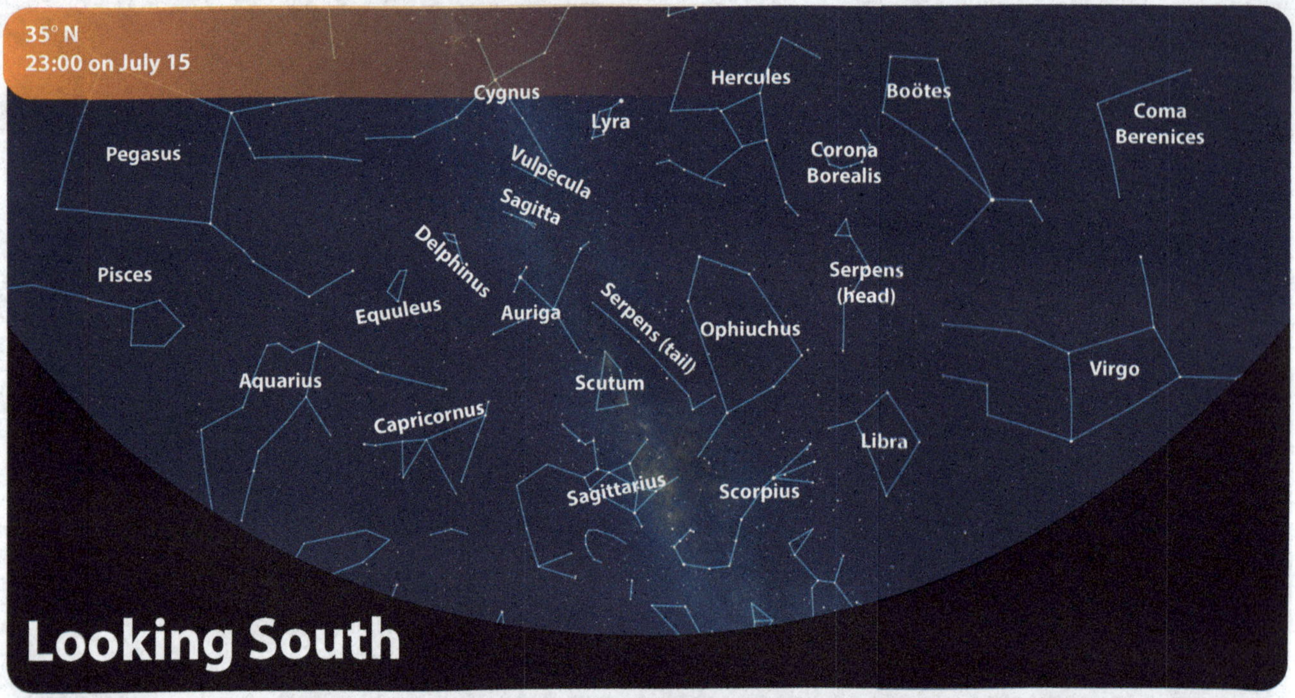

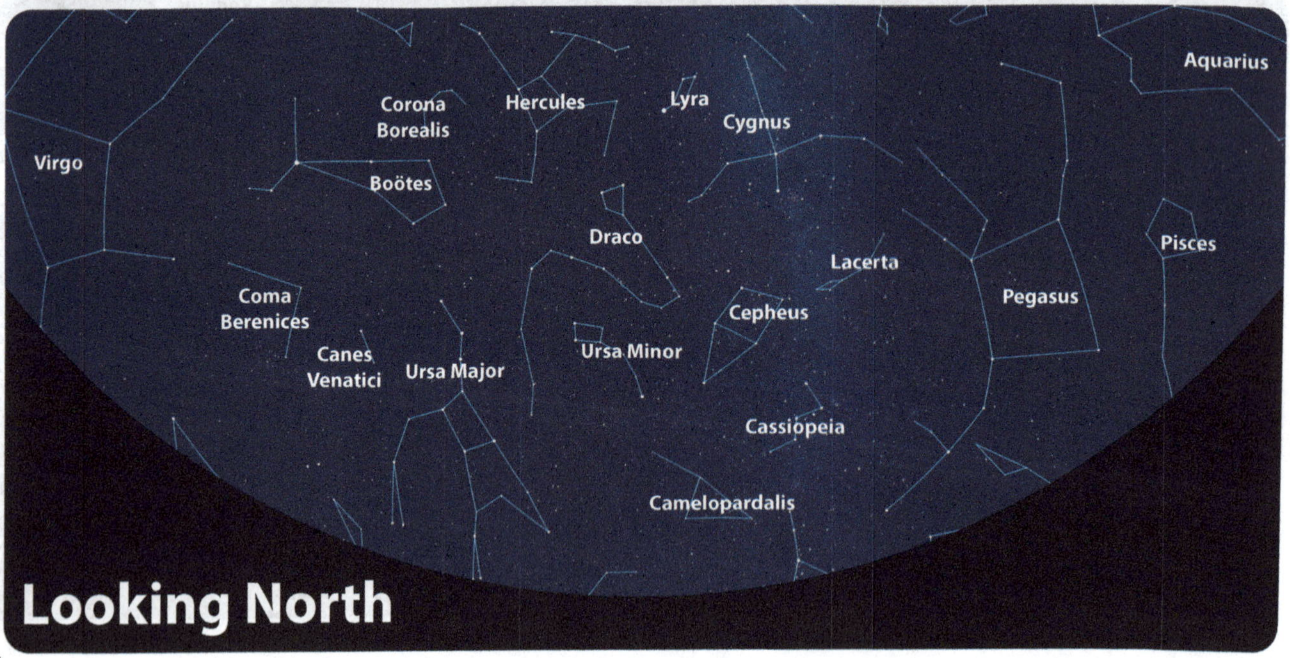

Autumn Full Sky Constellations

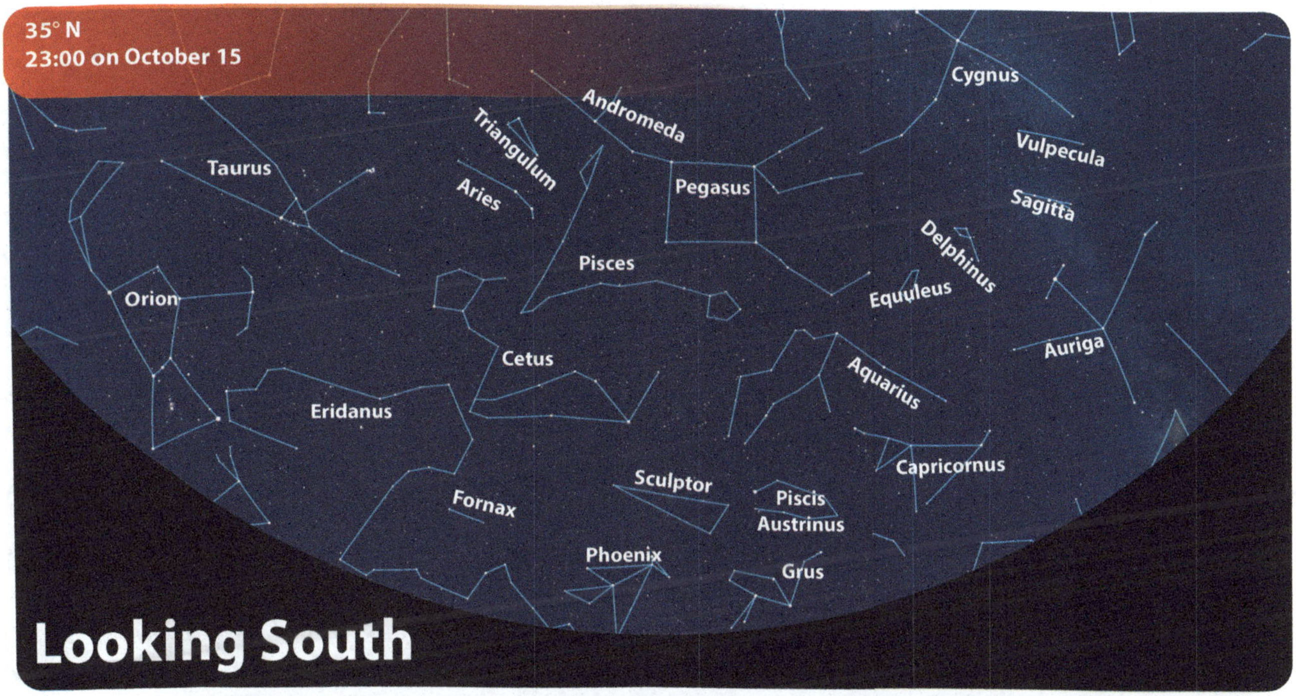

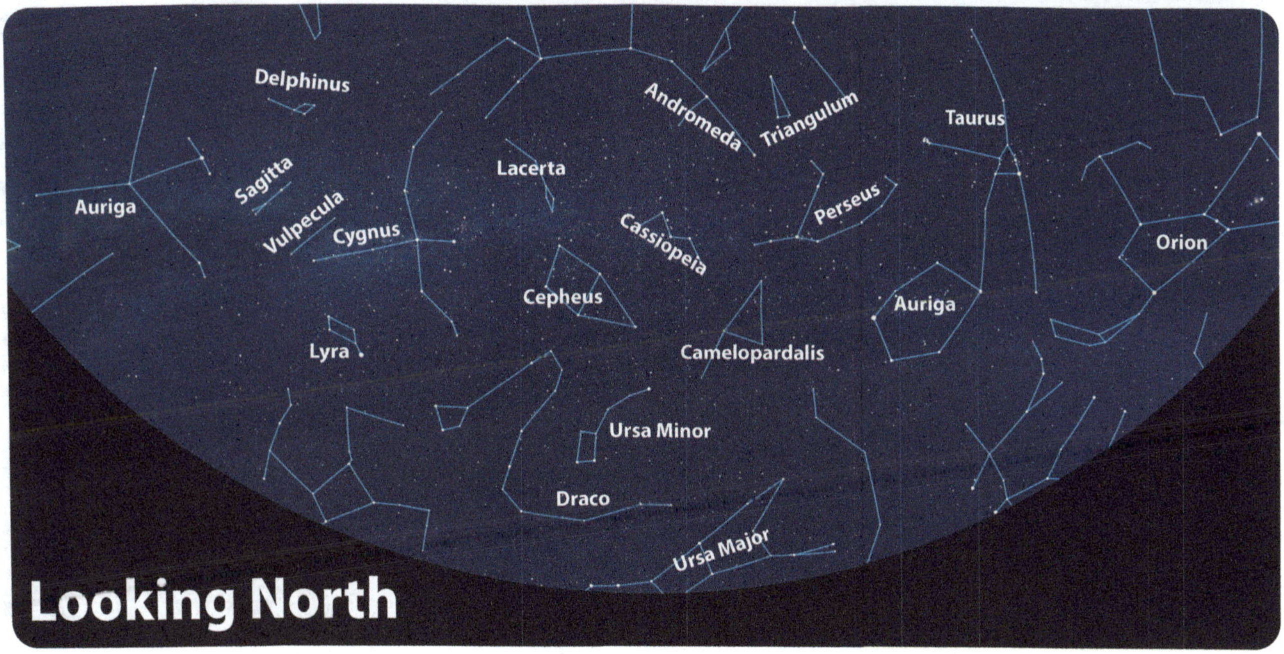

Winter Full Sky Constellations

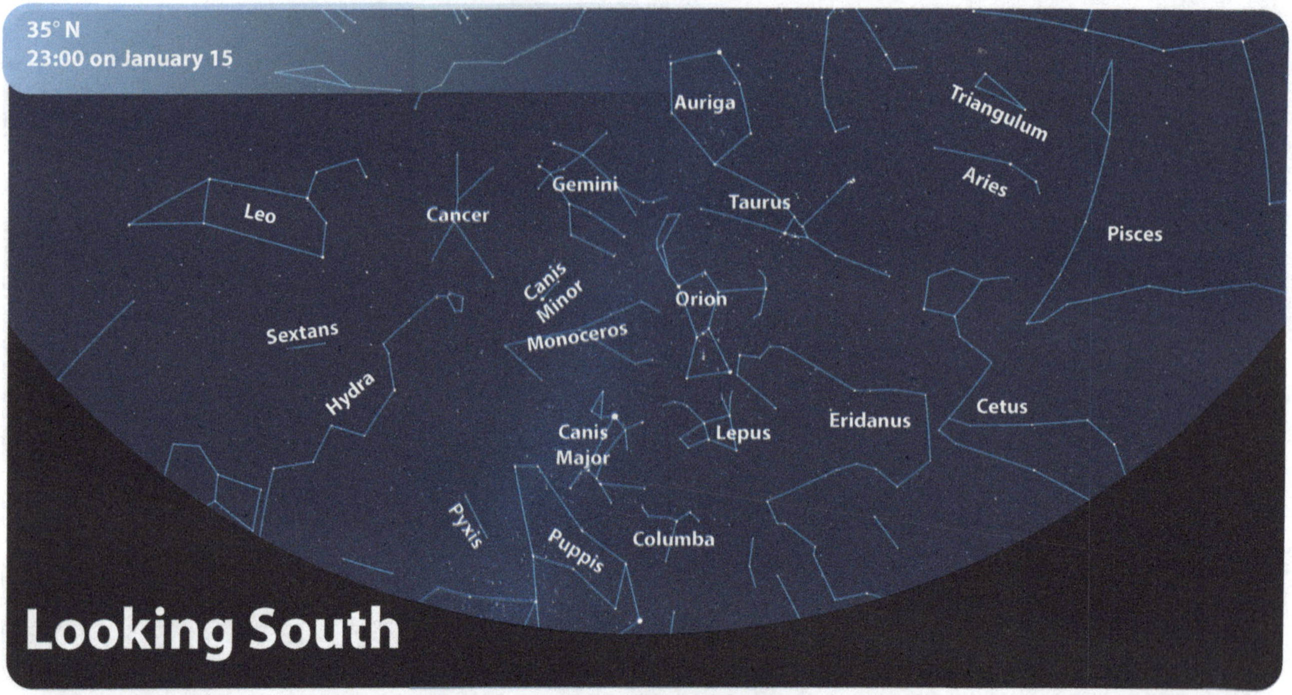

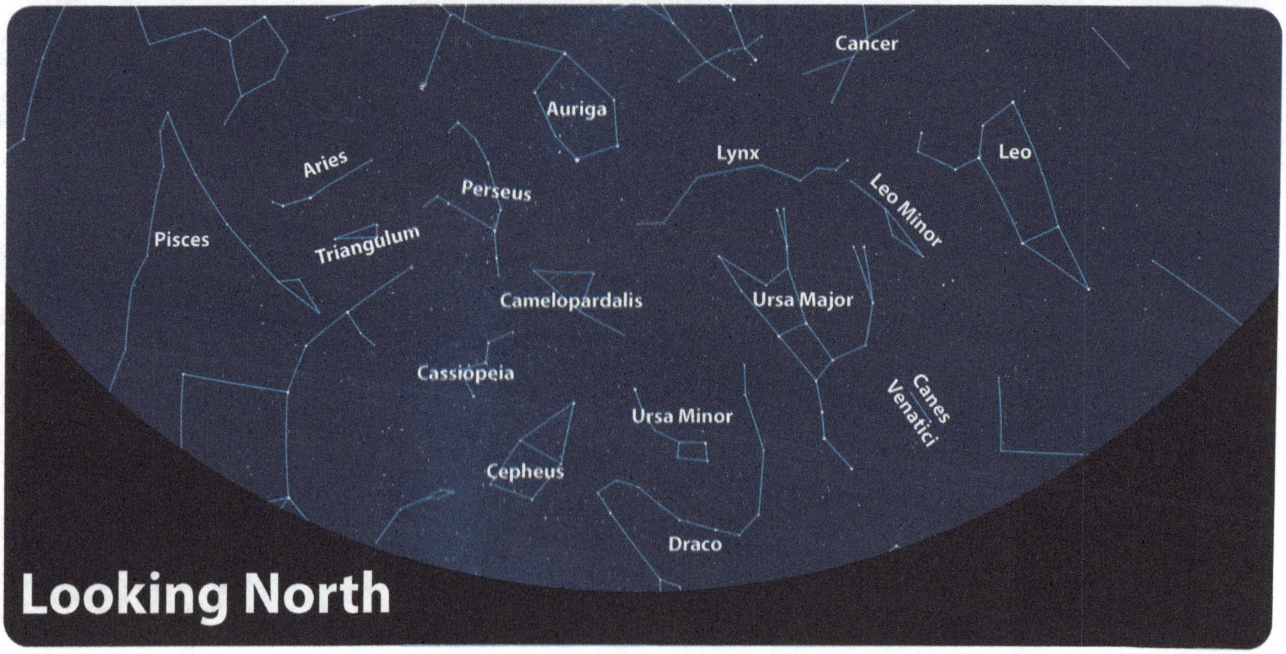

Spring Full Sky Constellations

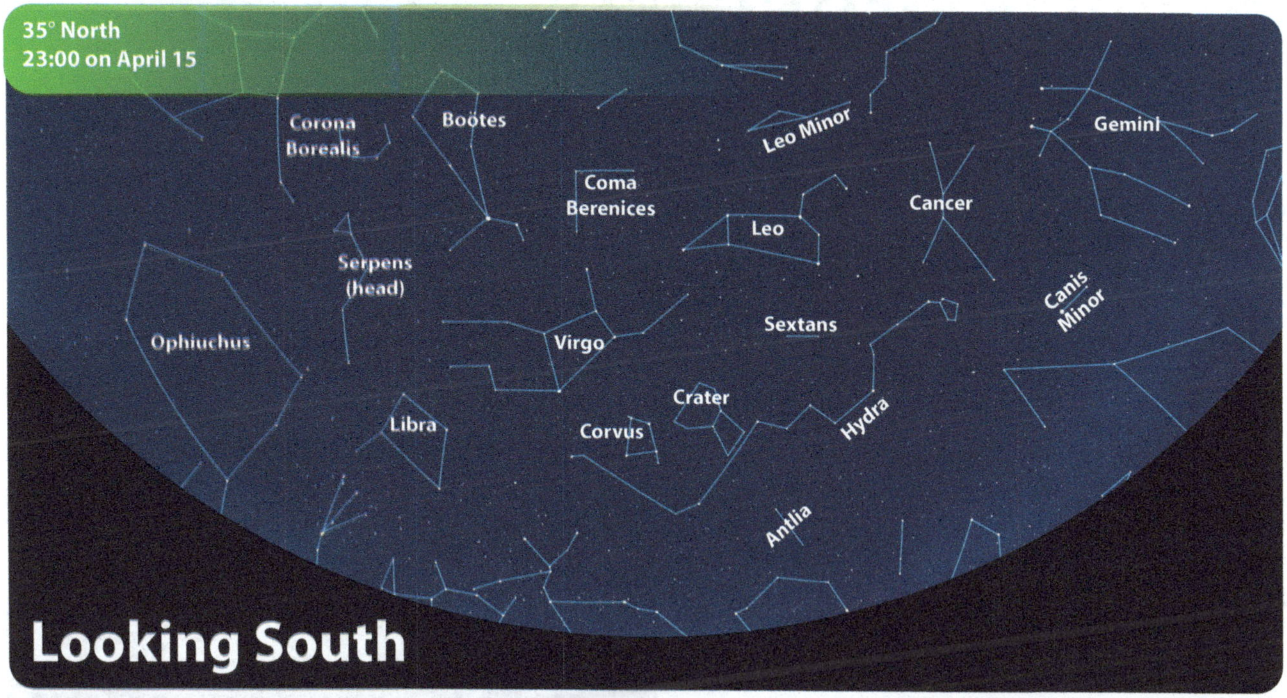

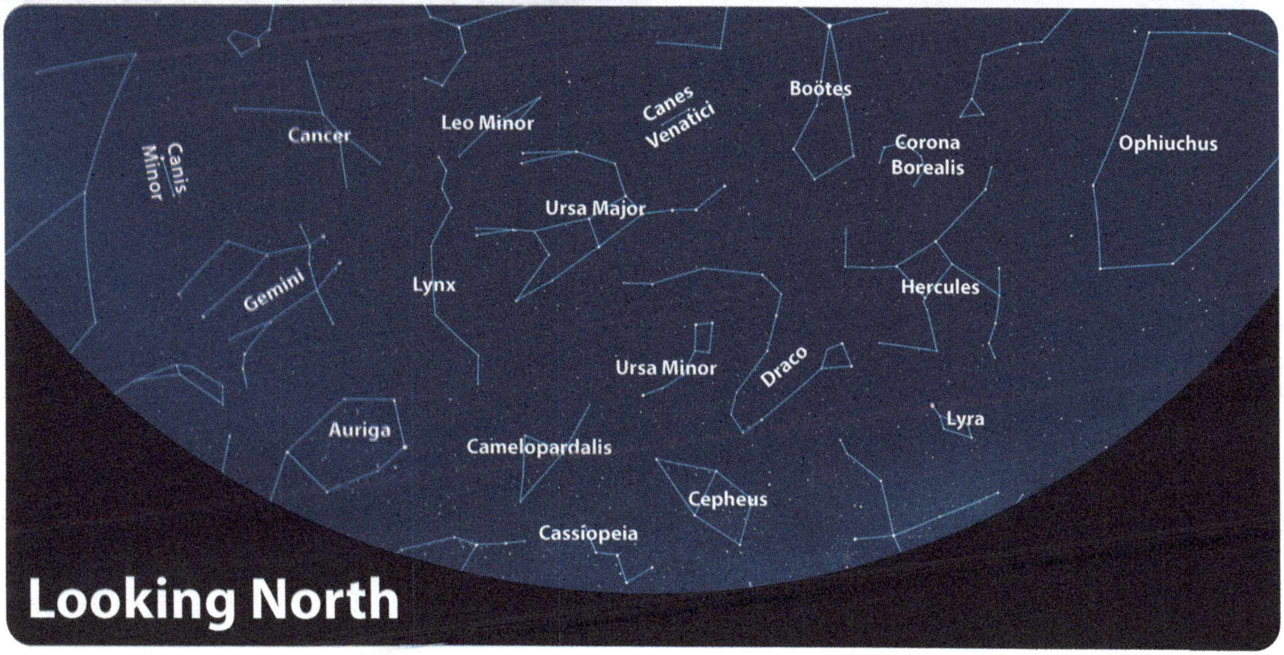

Beginner's Astrophotography

Capturing the Images for this Book

In the original version of *50 Things to See with a Small Telescope*, several of the photos were taken by friends of mine, and a few others, such as double stars, were simulated in Stellarium as it's actually quite difficult to photograph two bright stars next to each other (but easy to see with the unaided eye).

In this version of the book, I tried to capture most of the images myself, as I did in subsequent books. For *50 Things to See with Binoculars*, I used a 40-mm aperture telescope, which did a pretty good job capturing the images at a resolution comparable to binoculars. In this book, I've done the same thing, matching the aperture of the astrophotography rig to the aperture of the average beginner's telescope. An aperture of ~102 mm is on the larger end of what I would consider a "small" telescope, but it is the size I tend to recommend to beginners. So, that's what I used when capturing images for this book.

It's important to note that most of the images included in this book are **not** how typical astrophotographs would be presented. The nebulae, for example, have been processed to remove the colors picked up by the camera. I did this because this book is primarily about stargazing. I've done my best to edit the images such that the objects appear as they would in the eyepiece of a small telescope under ideal seeing conditions.

It should be noted that I am, first and foremost, a visual astronomer. I dabble in astrophotography as far as it serves me to capture the images shown in these books. I also try to maintain a beginner's mindset. While you may see me review astrophotography gear on my YouTube channel, *LearnToStargaze*, I do so as a beginner, and generally share my learning curve with the new equipment on camera (much to the chagrin of the experts in the comments section, who critique my use of the new gear to no end).

Most of the photos in this book were captured using a 103-mm refractor with a focal length of 700 mm, giving it a focal ratio of f/6.8. The telescope is apochromatic (APO), meaning that through a combination of lenses and/or high-quality glass, all wavelengths of visible light come into focus at the same point. APO telescopes eliminate chromatic aberration, a common issue with small beginner refractors with fast focal ratios. I included a 0.8× reducer field-flattener not really to reduce the focal length, but simply because that was what I used in my previous astrophotography setup and it seemed to work out okay.

Askar 103 APO refractor on the Skywatcher EQ6R Pro mount

The mount was an EQ6-R Pro, a bit oversized for the job. If I were picking up a mount in 2025 to use with this scope, I would have gone with the Sky-Watcher EQ-AL55i. One of these days, I'm going to injure myself carrying this (very heavy) setup across the lawn. The camera is an ASI2600MC Pro, a 26-megapixel cooled color camera. To capture an image with a similar field of view to that of the human eye at the eyepiece, you need a very large sensor. I believe this APS-C 23.5 mm × 15.7 mm sensor is probably the best in this regard. In the picture below left, you can see globular cluster M3 captured with this camera. Using the astronomy.tools website, I simulated the view through a standard eyepiece.

The images used in this book provide a "magnification" equivalent to about 30×. I used a ZWO ASIAIR to capture the data, live stacking mode to apply the flats, darks, and bias frames to the image in real time. Generally, I drop the images onto the computer, and use *Photoshop* to adjust the image so that it most accurately matches what you might expect to see in the eyepiece under ideal conditions.

Globular cluster M3 (compare with the simulated image at upper right)

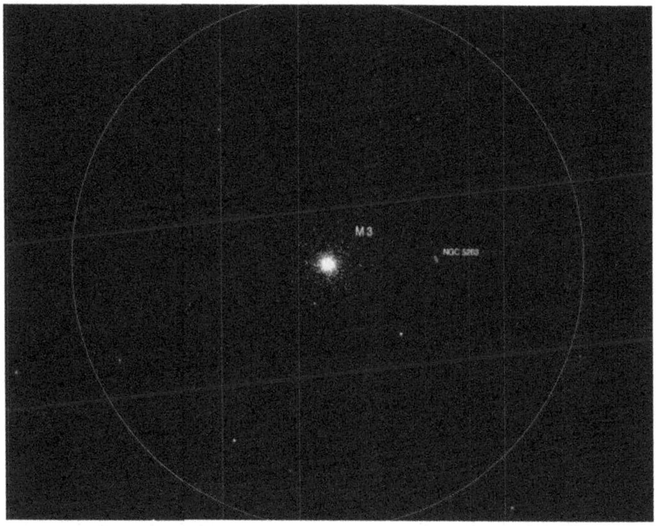

Simulated image of M3 as it would appear in a 28-mm, 56° field-of-view eyepiece with a 102-mm telescope at a focal length of 660 mm (24×)

Note that most galaxies appear extremely dim, but the photos show increased contrast simply so that they show up in print. Although photographs of galaxies and nebulae may often show color, your eyes will not see this. So, in most cases, I have removed any color from the images.

In instances where I was unable to show the feature I wanted to showcase with images taken with my backyard setup (like the Trapezium in Orion), I used images captured using the 24-inch telescope at the Burke-Gaffney Observatory at Saint Mary's University in Halifax, Nova Scotia, Canada, my hometown.

Even when using the observatory images, I still attempted to edit these in such a way as to be true to the backyard telescope experience.

Beginner's Targets

Astrophotography, although fun, is a very *different* hobby than stargazing. It's also far more challenging, more expensive, and yet arguably more addictive than other pastimes.

Astrophotography typically involves making every effort to bring out the detail in an object while removing as much noise as possible. The challenge with astrophotography is that targets like galaxies and nebulae are extremely dim and require incredibly long exposures to bring in enough data to create a detailed image.

This very short section only scratches the surface of astrophotography as a hobby. For each style of astrophotography, I've tried to articulate the easiest—not necessarily the cheapest—method for achieving the desired results.

The author shows a live view of the Moon using an ASIAIR and 127-mm Mak Telescope at a stargazing event in Oxnard California, July 2025

Milky Way Photography

Milky Way photography is the gateway to astrophotography. Setting a DSLR on a tripod, and capturing a 20-second exposure from dark skies at high ISO can provide an absolutely phenomenal image of the Milky Way. Even today's smart phones, set to "Raw" format, can capture impressive images.

To take this to the next level, a star tracker can be used to allow the camera to follow the stars, enabling much longer exposures without stars trailing in the image. Advanced photographers may use software to stack multiple images together to bring out even more details.

The Milky Way from Stargaze Nova Scotia

Lunar Photography

Lunar imaging is typically the first type of photography you'll do with a telescope. I've found that it's most easily done with a cell phone. You'll want to use a quality phone adapter like the Celestron NeXYZ or MSM Tridaptor. These are the only adapters that I know of that allow full control over how close the phone's lens is to the eyepiece.

Advanced lunar imaging techniques involve taking videos instead of pictures, and using software to combine the frames of the video into a single, high-resolution image.

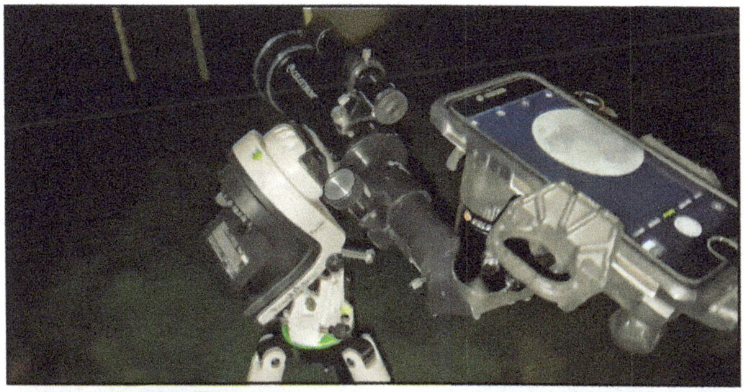

iPhone image of the Moon

Planetary Photography

Planetary photography can also be done with your phone. Typically, this is accomplished by taking a video of the planet—usually Jupiter or Saturn. The video is then transferred to the computer where each frame is "stacked" together into a single image. This technique is called "lucky imaging." In theory, the computer will select the best frames, and reject those of lower quality. The more frames in your video, the higher the quality of the image. On a PC, the software I use is PIPP to center the planet and convert the file to an AVI, and then Registax to stack the images. Several tutorials on YouTube go over this process in great detail.

The next level up from phone-camera imaging is by using a planetary imaging camera. These are designated astronomy cameras with tiny sensors. Planetary cameras can typically be found for around $200. They often come with software, such as ZWO's ASI studio, which can stack the images from the camera. More advanced (and free) software, such as FireCapture, can also do the job.

A basic color planetary imaging camera connected to a 200 mm Newtonian. A Barlow or coma corrector is required to reach focus.

Smart Telescope Photography

I've been fortunate to have tested most of the smart telescopes on the market today. It is challenging, however, to write any specific instructions for the use of a smart telescope, as each model is unique.

In this section, I'll attempt to summarize the current state of the smart telescopes available today (early 2026), in case you're looking to purchase one around the time of this book's publication.

The lowest cost smart telescope is currently the ZWO Seestar S30, priced at about $350. I found this extremely easy to use, and the images were impressive for a device this small. A pro version will be available soon with a much larger field of view.

In early 2026, the Dwarf Mini should arrive with a cost of about $400. This is even smaller than the S30 but seems to provide comparable images. One feature I enjoyed is their automatic image processor called Stellar Studio.

I've had the ZWO Seestar S50 for the longest. I've used it to capture, with ease, images of comets, galaxies, nebulae, and even the 2024 solar eclipse. Although the S50 is now an "older model," it receives frequent firmware updates, and new features are still being added.

The Dwarf3 seems to be based on the cameras used on the Mars rovers. Although its aperture is smaller at only 35mm, the images from the Dwarf3 seem to be on par with the S50 in terms of overall image quality.

The Vaonis Vespera II is what I'd like to call a "premium" smart telescope. This 50mm aperture telescope seems designed to replace a traditional astrophotography rig. Especially with Mosaic mode, some absolutely stunning images are possible, provided you spend enough time on target.

The Unistellar Odyssey Pro is really something special for a very particular user: the astronomy outreacher. The telescope includes an eyepiece experience where you look into an eyepiece, but you're actually looking at an internal screen containing the image being captured by the telescope in real time. What you see is a stacked and even processed image.

While stargazing traditionalists might find Unistellar's "Eyepiece Experience" a bit off-putting, as it seems a bit like cheating, the general public absolutely loves it. However, when compared to other smart telescopes, the images themselves currently don't hold up against the competition. If you're buying this primarily for the photographs, you may be disappointed.

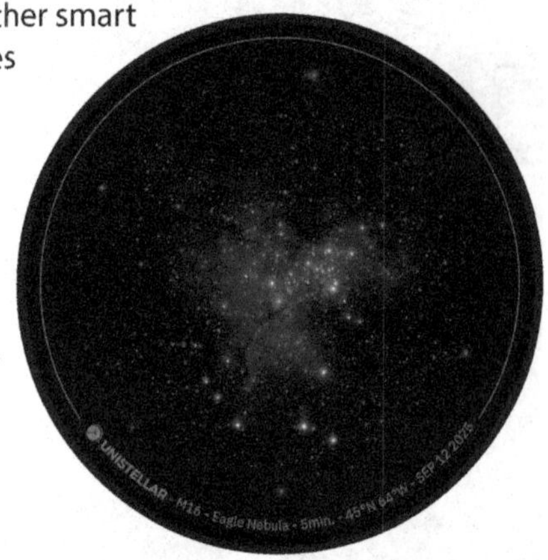

Image of M16 from Stargaze NS captured with the Unistellar Odyssey Pro. This is also an example of what you would see in this telescope's eyepiece experience.

Smart Telescope Best Practices

Stacking:

Even though all smart telescopes are different, the core principles are the same. They capture short exposures, typically between 10 and 60 seconds, and "stack" the images together. The more images the telescope stacks, the more detail you should expect to see (up to a point).

Although all smart telescopes can capture stunning images of nebulae, as well as decent images of the Moon and Sun (with a solar filter), none of the telescopes mentioned here are particularly good at imaging planets. This is primarily due to their limited apertures.

Exposure Time:

As with traditional astrophotography rigs, images of galaxies and nebulae don't tend to significantly overcome the background noise until after about one hour of exposure. The more light pollution you have, the longer it takes to cut through that noise.

When using mosaic mode (discussed below), much longer exposure times are recommended. Some telescopes offer a "multi-night" mode, as there is often not enough time in a single night to capture sufficient data.

EQ Mode:

The Dwarf and Seestar telescopes are now capable of EQ mode. If you're looking to capture high-quality images, this method is recommended. For quick captures—under 30 minutes, say—you might not notice a difference. However, for longer exposures, especially of dim nebulae, it makes a huge difference.

If you don't use EQ mode, field rotation starts to impact the images. This is where the object appears to rotate within the field of view, causing you to lose data in the corners of the images.

Mosaic Mode:

Most smart telescopes now feature a mosaic mode. Instead of being fixed to one spot in the sky, the telescope moves to capture a much wider field of view by tiling multiple images together.

Some smart telescopes, such as the Vaonis models, cannot be mounted equatorially. In this case, mosaic mode helps alleviate the loss of detail near the edges of your image.

Calibration:

All smart telescopes do some form of calibration, though some of this is transparent to the user. You may wonder why the telescope spends so much time "initializing," taking photos of random places in the sky, opening and closing the arm—that's all part of the process.

This is primarily to reduce the noise in the image, which we discuss in detail in the traditional astrophotography section of this book. Some calibration is manual; for example, Unistellar telescopes tend to need to be left outside for upwards of 30 minutes before use to help them adjust to the ambient temperature. Dwarf telescopes allow you to capture your own "dark" frames to further customize the image process.

Image Processing:

After the telescope has captured the image, the best results are often achieved by using a tool such as Photoshop to adjust the levels, color balance, and noise. Note that it's highly frowned upon in the astronomy community to manually add any details to the image that were not present in the original data.

Deep-Sky Astrophotography

As mentioned in the previous section, the methods described here may not be the cheapest or best way to capture these images. But, in my opinion, this is the absolute **easiest** way. For budget or extreme DIY astrophotography, see the brilliant Rory Griffin, and his Astrobiscuit YouTube and Discord Community.

DSLR with an Equitorially Mounted Telescope

A common first foray into deep sky astrophotography is to connect a DSLR camera to a telescope on an EQ mount. This allows you to take longer exposures while tracking the sky.

Using just a DSLR and telescope may limit you to only a minute or so of exposures before you start to see star trails in the image. In this case, the shorter the focal length of the telescope, the more manageable the system will be to use. The stars may also be out of focus at the edge of the field due to the lack of a field flattener. However, this is often a beginner's first experience with astrophotography, and it can be a LOT of fun.

One of my earliest attempts at astrophotography, connecting my DSLR directly to a refractor telescope on an EQ mount

Designated Astrophotography Rig

To reliably get amazing high-quality images of galaxies and nebulae, a designated astrophotography rig is typically required. This is where the hobby can become incredibly expensive. In general, when getting started in deep-sky astrophotography, small refractors are recommended. Newtonians can also capture amazing images, but the added complexity can be overwhelming (at least it was for me).

To avoid chromatic aberrations or other anomalies (such as bloated stars), the refractor telescope must be apochromatic (or APO). This ensures that all wavelengths of light have the same focal length. A designated field flattener is also required so that the stars appear sharp across the entire image (e.g., not elongated at the edges). The field flattener often serves as the adapter between the telescope and the camera, and as a place to attach any filters.

A basic deep sky astrophotography rig generally includes the following:

- A designated color deep-sky astronomy camera
- Guide camera and guide telescope
- Astrophotography-grade mount
- Field flattener
- Narrowband filter for nebulae
- Power Supply (usually a battery)

Advanced users typically add the following with time

- Mono-designated deep-sky astronomy camera
- Filter wheel
- Electronic focuser

These rigs are either controlled via a microcomputer, such as an ASIAIR, or connected to a laptop running software such as NINA (Nightscape Imaging 'N' Astronomy).

Basic Deep-Sky Astrophotography Rig

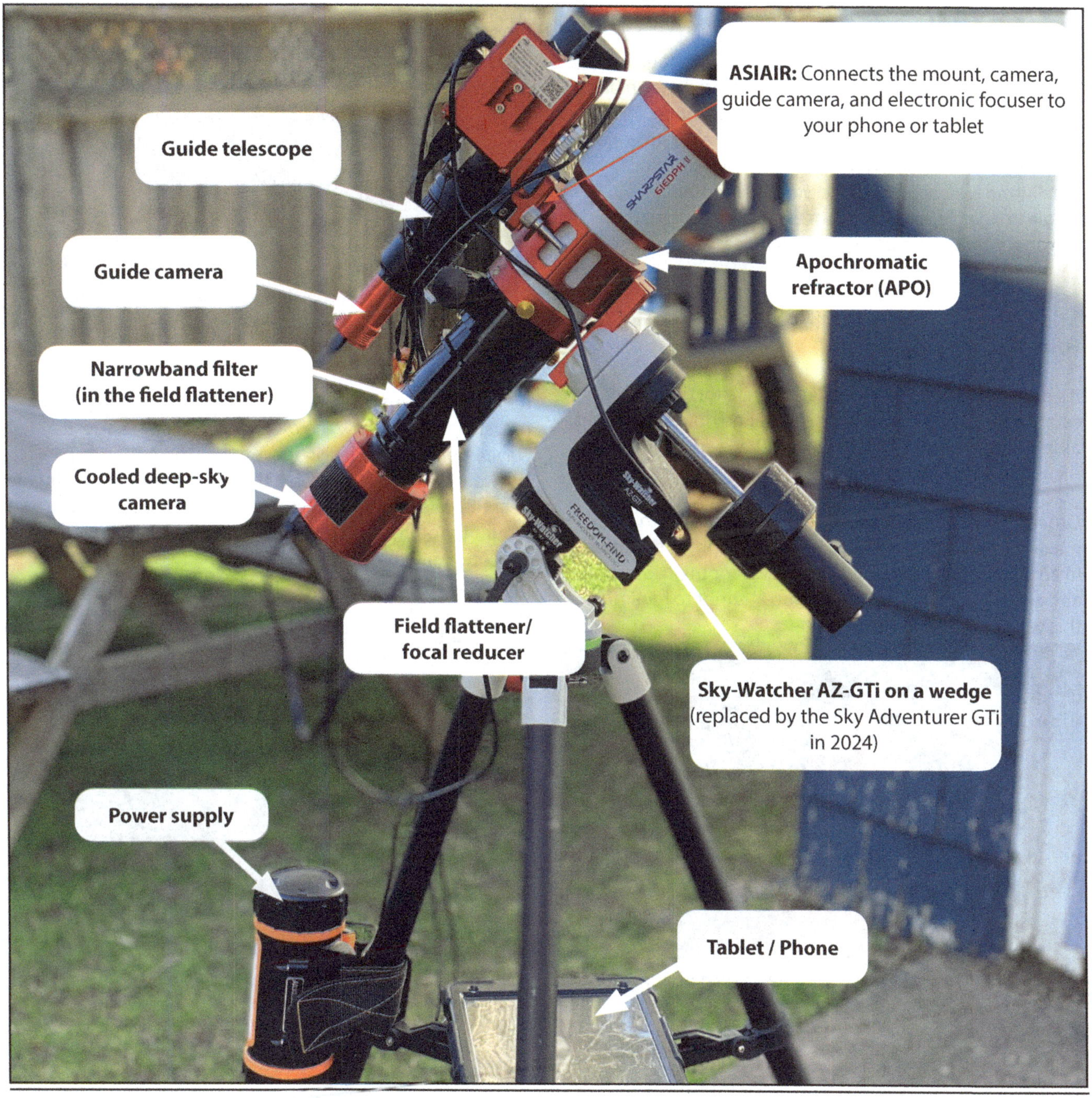

Capturing an Image

Even after you have all the gear, it's not as simple as hitting a button and capturing an image. The "data" for astrophotography must typically be gathered over many exposures, typically over a period of several hours (or nights). Each exposure must also be "calibrated" to remove unwanted noise inherent to the optical system itself. Finally, the exposures must be stacked together into a single image file, and then processed according to the astrophotographer's personal preferences.

Using an ASIAIR to zoom in on a star. The Bahtinov mask gives the star spikes such that when they are symmetrical (as shown), the telescope is in focus.

Focusing

It's extremely important to have accurate focus for astrophotography. Focus can also change during the night as temperatures change. If you don't have an electronic focuser, a Bhatinov mask is a reliable way to check the focus.

Other methods include using the FWHM (full width half maximum) features in your astrophotography application. In an ASIair for example, it gives a value for the size of a star, and as you twist the focus knob you attempt to minimize this value.

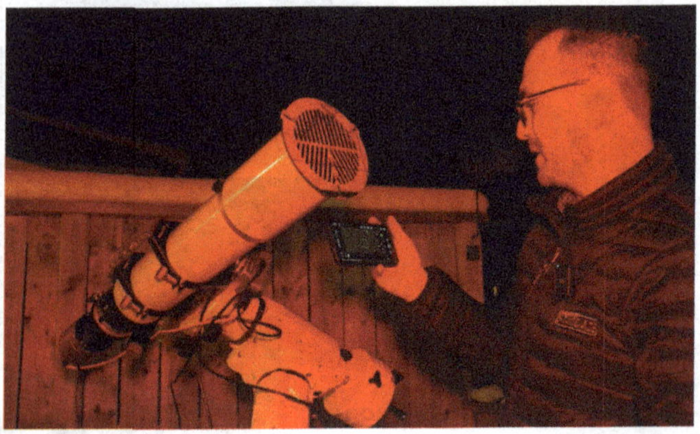

Focusing a telescope using a Bhatinov mask.

Polar Alignment

It's extremely important to have accurate polar alignment for astrophotography. The most effective (and easiest) way to polar align a telescope is to use the ASIAIR app, or software such as N.I.N.A. First, the camera captures an image near the celestial pole. Then, the computer instructs the mount to rotate by a certain amount. The system captures a second image, and by comparing the two, it calculates the required correction. Additional images are captured after each adjustment (of the mount's altitude and azimuth knobs) until sufficient polar alignment is achieved.

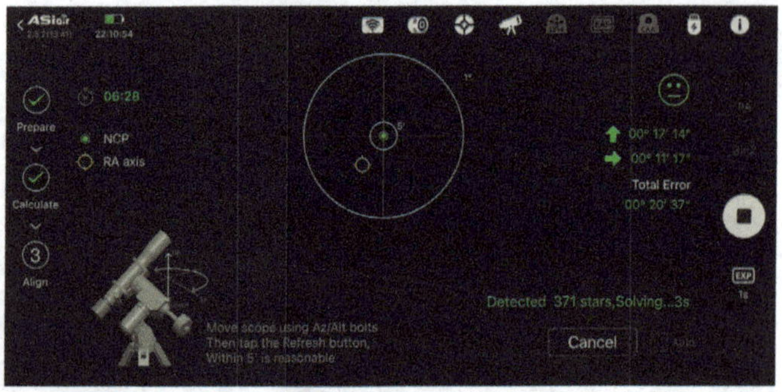

The ASIAIR's polar alignment process. Polar alignment is achieved by following the onscreen instructions. Adjustments are made until the error is reduced below a certain threshold.

Filters

"Narrowband" filters like the Optolong L-Enhance or L-Extreme vastly improve the quality of nebulae photos taken under light-polluted skies. These filters attempt to capture only the light from the nebula itself and block all other wavelengths, including those from the Moon and streetlights. Filters typically thread into the field flattener.

Narrowband filters are not recommended for galaxies, as they would block light from these objects. For imaging galaxies from the city, some astrophotographers prefer broadband filters like the L-Pro, which attempt to only block specific wavelengths (such as those produced by city lights), leaving a "broad" band of wavelengths available for the image.

From dark skies, the use of a filter while imaging nebulae is optional. And for galaxies, no filter is required.

Seasoned astrophotographers often switch to "mono" cameras. These are effectively simple "photon counters," which by themselves produce only grayscale images. They are typically used with a filter wheel and red, green, and blue (RGB) filters. Images exposed through these color filters must then be assigned that color during the processing phase of the astrophotographer's workflow.

A filter designed to be used with a color camera is threaded into the field flattener.

Live Stacking

"Live stacking" (a feature in apps such as ASIAIR) automatically applies the calibration images to the final image. This occurs in real time, as images (subs) are being captured, and previous images in the "stack" are often discarded immediately.

Using this method, there is only one image to keep track of. From a file management standpoint, I believe this is by far the simplest method to capture and store deep sky images.

More advanced users often use software on their computers to combine the calibration images with the light images. This process often requires a significant amount of hard drive space and *very* good file management.

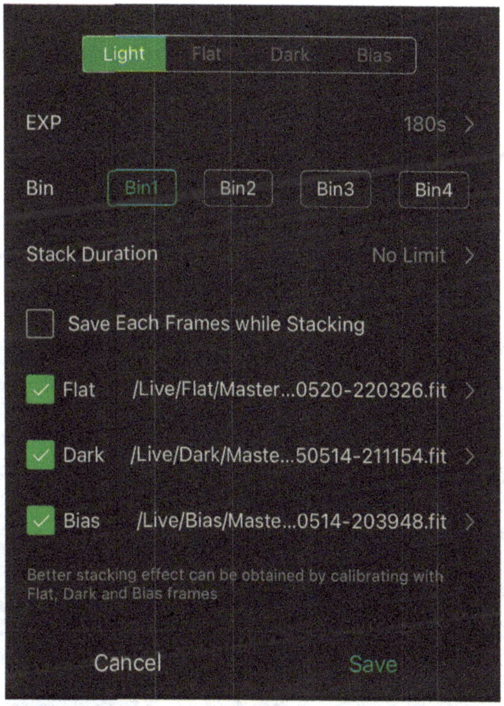

The Live Stacking menu in the ASIAIR app. These are the settings I typically use. Details on how to capture the flat, dark, and bias calibration images are covered on the next page.

Calibration Images

Because the deep-sky objects are so dim, astrophotographers need to know how to deal with "noise." Noise includes any artifacts that show up in your images, but shouldn't be there. A raw image includes several types of noise, including dust, heat from the camera's electronics (often called "amp glow"), and noise from the sensor itself. "Vignetting" is also a source of unwanted noise.

To overcome these issues, calibration images must be applied to the raw (aka "light") images. With ASIAIR or other apps, taking these calibration images is easy, and instructions are generally provided in the app itself. In summary, three types of calibration images that are recommended:

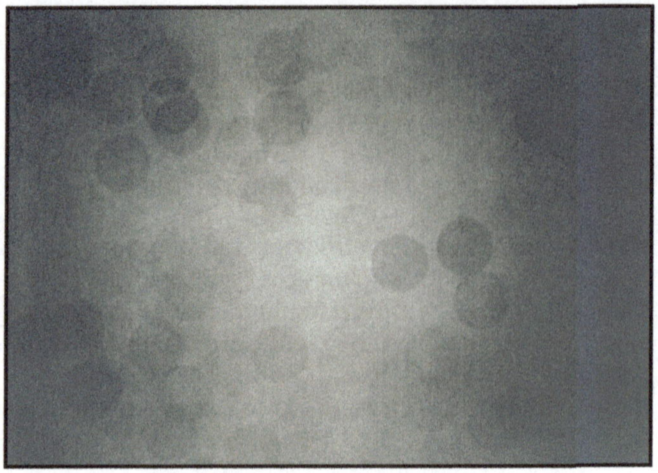

Your flats master file might look something like this (hopefully with fewer dust spots). The shaded circle is the vignette effect.

Flats:

Flats remove the vignetting effect and dust spots from the image. These are taken by pointing the telescope at an even light source, thus capturing an image of the unwanted artifacts. I use a $20 battery-powered LED tracing pad. Usually, about 20 flat images are taken, and then combined into a single "master" flat image. Flats should be taken every night as new dust can appear.

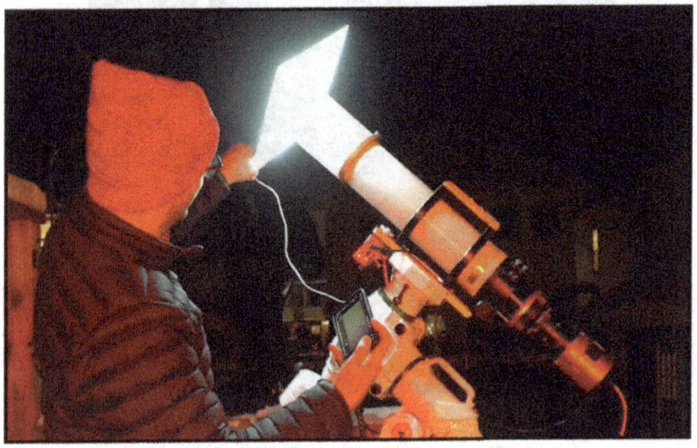

Taking flats (arguably the most important calibration image).

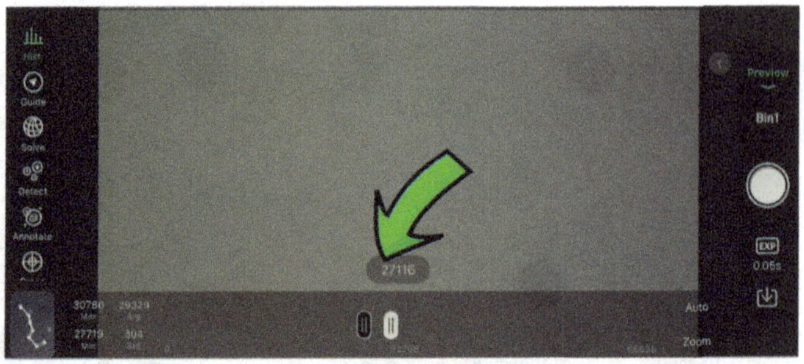

When taking flats, choose an exposure setting where the histogram peaks at about the halfway point, or a value of about 30,000 (ADU on a 16-bit sensor). For me, that's about 0.05 seconds.

Darks: Darks remove the noise generated by heat from the camera's electronics. These must be captured at the same exposure time and temperature as the "light" images, but with the lens cap on (hence, dark). Usually, about 15 darks are taken, and then combined into a master dark image. I usually retake my darks about once per season.

A dark master file might show significant amp glow. In newer cameras, this glow is significantly reduced.

Bias: Bias removes the noise from the sensor itself. These are captured at the same temperature setting, but on the absolute shortest exposure time (with the lens cap on). Usually, about 40 bias images are captured and combined into a single master bias image. I retake my bias images as often as I capture my darks. Bias images basically look like flat black images, but may include bright spots representing "hot pixels."

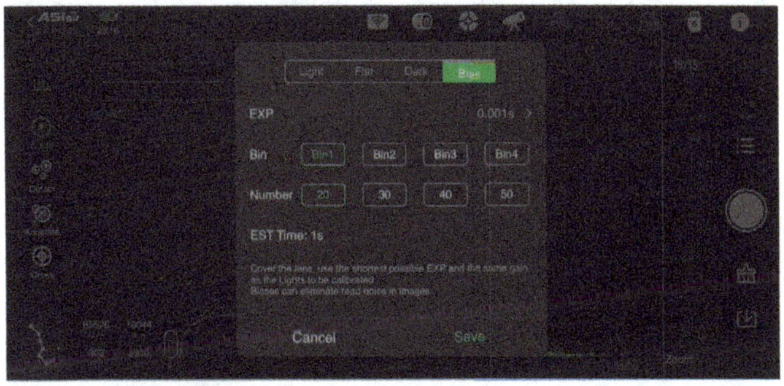

"Bias" settings within the ASIAIR's live stack menu.

Bin: Binning is the process of treating two or more pixels as a single pixel. I recommend keeping this set to Bin 1.

Guiding

An autoguiding system allows your telescope to capture long individual exposures while minimizing the risk of tracking errors, such as star trails. While PC users typically rely on software called "Push Here Dummy" (PHD2), the ASIAIR method is far easier. Guiding with an ASIAIR is accomplished directly through the "Guide" menu. From there, you can start capturing images from your guide camera. Keep in mind that the autoguider does not need to be perfectly aligned with the main telescope like a finder scope would, but it absolutely must be focused. The first time you use the autoguider, or if you have made any physical adjustments to the setup, the software will need to be calibrated.

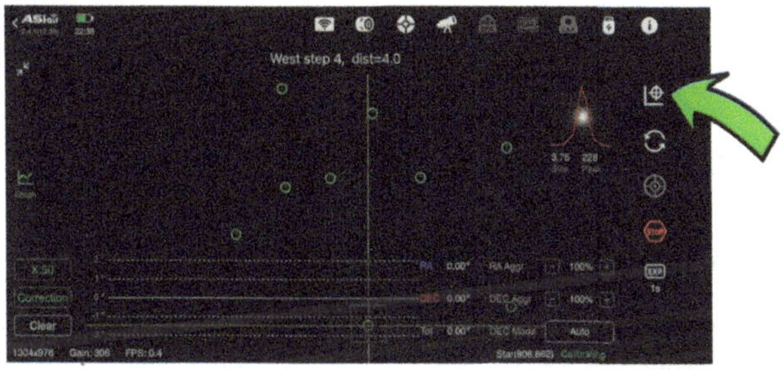

The ASIAIR's "Guide" menu. The green arrow points to the Calibration button.

Total Exposure Time

There are many schools of thought regarding how much exposure time is necessary to overcome the noise, and bring out the fine details in your target. In any case, the minimum total exposure time is generally considered to be one hour under any conditions. Each individual image (often called a "sub") is typically obtained with 60 seconds to 5 minutes of exposure time—not so short as to generate excessive files, but not too long as to overexpose the image (I use 3 minutes for almost all deep sky targets).

That said, the darkness of your sky matters a lot. In a very dark sky, the camera will gather an astonishing level of detail in a very short time. In the city, however, one hour of exposure might be satisfactory, but leave you wanting more. Seasoned astrophotographers typically image a single target for 5–10 hours or more, and often over several nights.

Processing a RAW or FIT Image

With the calibration and stacking complete, it's time to process the image. Just like a regular photographer adjusts a photo to get the colors "just right," astrophotographers must process their data in order to bring out all the details in the object they have imaged.

Astrophotography is far too broad a subject to cover completely in a beginner stargazing book such as this. There are dozens of pieces of software, each of which would require an entire book to go over in any detail. That said, the core concepts you need to understand in order to process an image (from a designated astronomy camera) are presented below.

File Types

Image processing on the computer typically involves "FIT" files, which are designed to be opened in specialized astronomy software. If you want to process the image in *Photoshop*, the file will need to be converted to a 16-bit TIFF file. Recently, I've been using software called *GraXpert* to accomplish this task.

After you've finished processing the image, you might consider "compressing" the image to a PNG (lossless compression) or JPG (lossy compression).

Processing Software:

ASIAIR + device: If you're using an ASIAIR or similar device, it's possible to save and process the image on the device itself using the app of your choice.

Siril: Siril is a free and open-source astrophotography processing software. This is good for beginners to handle basic processing.

GraXpert: Free and open-source software designed for only three tasks: remove gradients, denoise, and export a file to TIFF format so it can be processed in *Photoshop*.

Photoshop: *Photoshop* is a popular and powerful image processing tool. It can also be customized with specialized tools (called actions) to perform astrophotography-specific tasks. This is what I used to process the images captured for this book.

GIMP: A free and open-source, but somewhat limited, alternative to *Photoshop*. Note: *GIMP* can open FIT files.

PixInsight: Advanced users might use software such as *PixInsight* to process their images to taste. However, this software is expensive, and has an incredibly steep learning curve. If you're interested in going this route, there are several YouTube videos that would be far more effective than a using beginner stargazing book. If you are interested, I recommend the YouTube Channel *Nebula Photos*.

Gradient Removal

If you collected the data from light-polluted skies or when the Moon is out, it's likely your image will have a gradient. That's when one side of the image is brighter than the other. *GraXpert* is a great tool for removing this issue, ensuring that the background sky appears uniform. in your image.

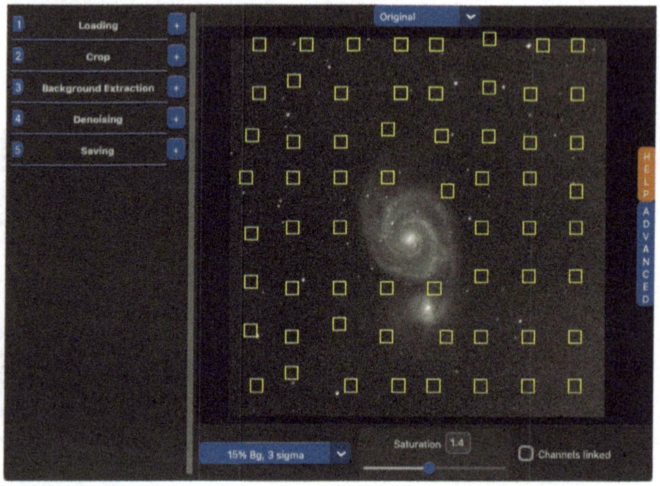

Using GraXpert to remove the gradient from an image of M51

Color Balance

Sometimes, even if you're using a color camera, the image may appear too red. This is because the sensor is more sensitive to red light. Like stretching, some programs, like *GraXpert*, will attempt to do this for you. In *Photoshop*, the colors can be balanced to taste by adjusting the channel (red, green, or blue) in either the "levels" or "curves" menus. More of an art than a science, several videos on YouTube can guide you in this process.

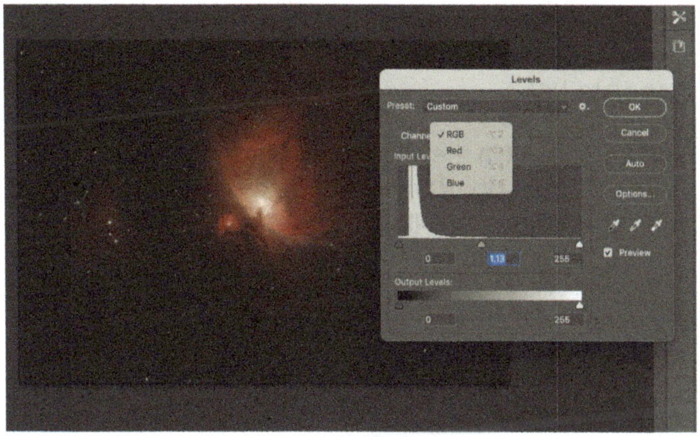

Adjusting the color balance by selecting each color channel in the "levels" window.

Stretching an Image

Even after the "final" image is calibrated, sometimes opening the image file results in a black screen with only a few stars showing at all. This is totally normal. The camera has, in fact, simply gathered an image of mostly darkness. Don't worry, the nebula is in there, we just need to differentiate between the various *levels* of darkness. In other words, we need to turn up the brightness on the slightly less dark parts of the image!

This process of making the least dark parts brighter while keeping the darkest dark parts dark is called "stretching" the image. This can be done in several ways. Some software, like *GraXpert*, can do this for you. In *Photoshop* this can be done either in "levels" or "curves." Similar to color balancing, there's no perfect way to do this, and you will want to use your own taste (and YouTube tutorials) to guide you through the process.

Extremely Basic Imaging Workflow

Some astrophotographers love processing and spend hours working on a single image. My standards are not quite that high. I'm typically satisfied by completing the processing in a few relatively simple steps without specialized software. Here is an example of how an image might be processed in *Photoshop*, starting with the fully calibrated 16-bit TIFF file:

Step 1: Adjust levels and curves (in an adjustment layer) until the object and the stars become visible. No part of the image should be over or underexposed.

Step 2: Adjust the color channels (in the layers or curves menus) and tint so that the galaxy or nebula looks as it should and the background looks like, well, the sky.

Step 3: In "Camera Raw Filter" menu, use the clarity, shadows and highlights sliders to bring out any additional detail.

Step 4: Also in "Camera Raw Filter," apply noise reduction to the image such that the background appears smooth and sharp, but not too much or the image will start to look like an oil painting.

Step 5: Save as a TIFF (uncompressed) or PNG (compressed).

As I've said throughout this section, this is extremely basic. YouTube is, by far, the best place to get detailed step-by-step instructions on how to process your astrophotos.

Glossary of Astronomical Terms

Asterism: An unofficial name for a grouping of stars. Used to identify patterns in the night sky. The most popular asterism is the Big Dipper within the constellation Ursa Major.

Averted vision: A technique for looking at deep-sky objects where a dim object brightens when you focus your gaze beside it instead of directly at it.

Azimuth: Refers to the left-right movement of a telescope and a corresponding direction or location in the sky expressed in compass degrees.

Bortle Scale: A measure of sky brightness ranging from 1 (extremely dark, rural skies) to 9 (bright city skies).

Conjunction: When two objects, such as the Moon and a planet, meet or pass each other in the sky.

Constellation: A region of the sky containing recognizable patterns of stars. The sky is divided into 88 of these regions.

Declination: The angular distance from the Celestial Equator.

Deep-sky object: A stargazing target that resides outside our Solar System, including galaxies, nebulae, and star clusters.

Ecliptic: The apparent path the Sun takes against the background stars throughout the year. The Moon and planets are always found near the Ecliptic.

Light year: The distance light travels in one year. It is used to measure distances in space. One light year is about 9,500,000,000,000 kilometres (~10 Trillion).

Magnitude: A numerical scale that describes the apparent brightness of an object in space as observed from Earth. Higher magnitudes denote dimmer stars.

Meridian: An imaginary line beginning at the southern horizon, passing through the zenith, and terminating at the northern horizon. Separates the eastern and western parts of the sky. Objects rising in the east reach their maximum altitude at the Meridian.

Messier (M) list: A list of 110 objects that originated with French comet hunter Charles Messier about 200 years ago.

New General Catalogue (NGC): A list of almost 8,000 nebulae and star clusters originally compiled by John Dreyer in the late 1800s.

Node: The point in the Moon's orbit where it crosses the Ecliptic. Eclipses occur when the Moon passes this point during New Moon or Full Moon.

Orbit: The curved path an object, such as a planet, moon, or spacecraft, takes as it travels through space.

Opposition: When a planet is directly on the opposite side of Earth from the Sun.

Right Ascension: The distance east and west in the sky expressed in hours, minutes, and seconds. Zero is where the Sun crosses the celestial equator on the first day of spring.

Sidereal: Time and motion measured relative to distant (fixed) stars. Used by motorized mounts to track stars and deep sky objects.

Sunspot: A magnetic storm on the surface of the Sun that is visible in binoculars or a small telescope.

Waning: Refers to the Moon's phases as it moves from Full Moon back to New Moon.

Waxing: The Moon's phases between New Moon and Full Moon. These are regarded as the best phases to observe the Moon as it is well positioned in the evening sky.

Zenith: The point in the sky directly above the observer.

Observing Certificate Programs (Free)

Option 1

If you have observed most of the objects in this book and recorded your observations as you went along, it's likely you have met all the requirements for the RASC's Explore the Universe (ETU) Program. See the attached website for the requirements.

You do NOT have to be a member of the RASC or live in Canada to apply for your ETU certificate. Use the application form below.

https://www.rasc.ca/explore-universe

Option 2

If you are a member of an astronomy club in the USA, and a member of the Astronomical League, you may apply for the Universe Sampler Observing Program certificate and pin.

To receive your certification, an approved club member must validate your observing logs (such as those found in this book) and email the Universe Sampler Observing Program Coordinator to request your certificate and pin. Details found here:

astroleague.org/universe-sampler-observing-program

RASC Explore the Universe Observing Program Application for Certificate

Congratulations on completing the RASC Explore the Universe Observing Program! To receive your certificate, please complete this form and submit it as follows:

for RASC Members of Local Certification Centres	for National Members, Non-RASC Observers, or Members of National Certification Centres
If your RASC Centre is a Local Certification Centre of the Explore the Universe Observing Program, submit this application form to a representative of your local Observing group or a member of your local Executive. Consult your local RASC website or newsletter for more details. Scanned files may be shared with the *Uploader*.	Forward your completed application form to: Attn: Observing Committee Royal Astronomical of Society of Canada 203-489 COLLEGE ST TORONTO ON M6G 1A5 Email observing@rasc.ca Do not email attachments. All applicants or reviewers, ask for the *Uploader Service* link to share files, logs, sketches, and photos.

Affidavit

I, ______________________, attest to the following:

1. I have found and described all of the objects listed below myself.
2. I have not used a computerized "go-to" telescope to meet the requirements outlined in the program.
3. I have completed the required number of observations for each section of the program.

Name ______________________ Date ________ (yyyy-mm-dd)

Address ______________________

City, Code ______________________

E-mail ______________________

Certificate Language ⦿ English ◯ French (indicate one)

Authentication

(Authentication will be performed by a qualified member of your Centre or by the Observing Committee representative.)

I declare that this form has been filled out completely and correctly and that the individual observer named to the left has qualified for the *Explore the Universe Certificate* of The Royal Astronomical Society of Canada.

Name ______________________ Date ________ (yyyy-mm-dd)

RASC Centre ______________________

Observing Log

Date:	Time:	Location:

Sky Conditions (Seeing/Transparency):

Instrument (Magnification/Aperture):

Notes:

Observing Log

Date:	Time:	Location:

Sky Conditions (Seeing/Transparency):

Instrument (Magnification/Aperture):

Notes:

Observing Log

Date:	Time:	Location:

Sky Conditions (Seeing/Transparency):

Instrument (Magnification/Aperture):

Notes:

Observing Log

Date:	Time:	Location:

Sky Conditions (Seeing/Transparency):

Instrument (Magnification/Aperture):

Notes:

Photocopy this page for extra observing logs

Observing Log

Date:	Time:	Location:

Sky Conditions (Seeing/Transparency):

Instrument (Magnification/Aperture):

Notes:

Observing Log

Date:	Time:	Location:

Sky Conditions (Seeing/Transparency):

Instrument (Magnification/Aperture):

Notes:

Observing Log

Date:	Time:	Location:

Sky Conditions (Seeing/Transparency):

Instrument (Magnification/Aperture):

Notes:

Observing Log

Date:	Time:	Location:

Sky Conditions (Seeing/Transparency):

Instrument (Magnification/Aperture):

Notes:

Target Index

Photo Credits

All photos provided by the author unless otherwise stated
All star maps from Stellarium
Photos of Smart Telescopes are from the manufacturer
Burke-Gaffney Observatory (BGO)

15: Michael A. Phillips
20: (Right-Bottom) BGO
21: (Left) BGO, (Right-Bottom) BGO
23: (Left-Bottom) Tim Doucette
50: Blair MacDonald
70: BGO Halifax, Nova Scotia, Jul 21, 2019
72: BGO Halifax, Nova Scotia, Aug 28, 2018
85: Dr. David Hoskin
89: BGO
99: (Right) BGO
107: BGO Halifax, Nova Scotia, December 17, 2017. 40-second exposure by John Read
108: BGO
109: WorldWide Telescope
112: (Left and Right) David Hoskin
113: (Left, Middle, and Right) BGO
114: Stellarium
116: (Top-Left and Bottom-Left) BGO
118: BGO Halifax, Nova Scotia, Jan 9, 2017. 20-second exposure by John Read
121: Blair MacDonald
123: Blair MacDonald
125: (Bottom) Digital Sky Survey
128: (Right) Digital Sky Survey

Photo Details

8: Graves Island, Nova Scotia. Canon t3i DSLR.
9: (Left) The Milky Way. Sony A7siii Camera, 15-second exposure. Abbey Ridge Observatory, Stargaze Nova Scotia (NS).
10: Sony A7s, focal length 28 mm, 20-second exposure. Stargaze NS. December 14, 2023.
11: Sony A7siii, focal length 28 mm, 10-second exposure. Abbey Ridge Observatory, Stargaze NS. Oct 11, 2024.
12: (Top) Canon T3i, focal length 50 mm, 0.25-second exposure. Halifax, NS, Nov 8, 2022.
13: Sony A7siii (set to auto). Apr 8, 2024.
22: (Right) SeeStar S50 smart telescope. Oct 18, 2024.
42: (Right) iPhone though a small telescope.
43: Sony A7s, focal length 28 mm, 20-second exposure. Stargaze NS, Mar 12, 2023.
52: Venus. 90-mm telescope.
58: Seestar S50. Doaktown, New Brunswick, Apr 8, 2024.
59: Seestar S50, focal length 250 mm, aperture 50 mm, best 69 exposures of a 4-second video. Halifax, NS, Jun 14, 2024 (black border added).
61: Sony A7siii, 20-second exposure. Jun 29, 2024.
62: 102-mm refractor, Canon DSLR, 30-second exposure. Halifax, NS, May 7, 2017.
63: CF 102, 294MC, 10-second exposure. Halifax, NS, May 18, 2022.
64: Askar 180, 294mc Pro, 6-minute exposure, Kejimkujik, Aug 12, 2023.
65: Askar 103, 2600MC, 4-minute exposure. Stargaze NS, Aug 31, 2025.
66: (Left) 24inch telescope, (Right) Askar 103, 2600MC, 6-minute exposure. Stargaze NS, Aug 31, 2025.
68: Askar 180, 2600MC, 6-minute exposure, Tucson, May 10, 2024.
69: Askar 103, 2600MC, 24-minute exposure, Halifax, NS Feb 10th, 2026
71: Askar FMA180, 294mc Pro, 6-minute exposure, Kejimkujik, Aug 12, 2023.
74: Askar 180 (40-mm aperture), 2600MC, 9-minute exposure, Tucson, May 9, 2024.
76: Askar 103, 2600MC, 3-minute exposure. Stargaze NS, Jun 22, 2025.
77: Sharpstar 61, 294MC, 3-minute exposure. Stargaze NS, Aug 23, 2023.
78: Askar 103, 2600MC, 6-minute exposure. Stargaze NS, Jun 22 2025.
79: Askar 103, 2600MC, 15-minute exposure. Stargaze NS, Jun 22 2025.
80: Askar 180, 294MC, 6-minute exposure. Kejimkujik, Aug 11, 2023.
81: Askar 103, 2600MC, 6-minute exposure. Stargaze NS, Jun 22, 2025.
82: Askar 103, 2600MC, 6-minute exposure. Stargaze NS, Jun 22, 2025.
83: CF 102, DSLR, 35-second exposure. Smileys Provincial Park, Jul 20, 2017.
84: Askar 103, 2600MC, 9-minute exposure. Stargaze NS, Jun 22, 2025.
86: Askar 103, 2600MC, 9-minute exposure. Stargaze NS, Jun 22, 2025.
87: Askar 103, 2600MC, 9-minute exposure. Stargaze NS, Jun 22, 2025.
88: Askar 103, 2600MC, 3-minute exposure. Stargaze NS, Aug 31, 2025.
90: Sharpstar 61, 294MC, 6-minute exposure. Halifax, NS, Aug 10, 2023.
91: Askar 103, 2600MC, 6-minute exposure. Stargaze NS, Jun 21, 2025.
92: (Left) Sharpstar 61, 294MC, 6-minute exposure. Halifax, NS, Aug 10, 2023 (Through Haze).
92: (Middle) Askar 103, 2600MC, 3-minute exposure. Stargaze NS, Aug 31, 2025.
94: Askar FMA180, 2600MC, 9-minute exposure. Stargaze NS, Aug 2, 2024.
97: Askar 103, 2600MC, 60-minute exposure, Halifax, NS Feb 10th, 2026
98: Askar 103, 2600MC, 16-minute exposure. Stargaze NS, Aug 31, 2025.
99: (Left) Z100 $99 Telescope (on a Tracking Mount)
100: Askar 103, 2600MC, 3-minute exposure. Stargaze NS, Aug 31, 2025.
101: Askar 180, 294MC, 6-minute exposure. Halifax, NS, Oct 24, 2023.
102: (Left) Askar 103, 2600MC, 1-minute exposure. Stargaze NS, Aug 31, 2025.
102: (Right) Askar 103, 2600MC, 1-minute exposure. Stargaze NS, Aug 31, 2025.
104: Askar FMA 180. Halifax, NS, Oct 23, 2023.
106: Orion Nebula. SharpStar 61-mm 294 MC Pro Camera, Halifax, NS.
110: Askar 180, 294MC, 9-minute exposure. Halifax, NS, Oct 18, 2023.
116: (Right) Askar 180, 2600MC, 3-minute exposure. Stargaze NS, May 25, 2024. (Bottom-Middle) Askar 180, 294MC, 3-minute exposure. Halifax, NS, Oct 3, 2023.
120: Askcar 180, 2600MC, 12-minute exposure. Stargaze NS, May 25, 2024.
122: Askar 103, 2600MC, 12-minute exposure. Stargaze NS, Jun 21, 2025.
124: Askar 180, 2600MC, 12-minute exposure. Stargaze NS, May 25, 2024.
125: (Top) Askar 180, 2600MC, 6-minute exposure. Biosphere 2, Arizona, May 10, 2024.
128: (Left) Askar 180, 2600MC, 9-minute exposure. Halifax NS, May 24, 2024.
128: (Middle) Askar 180, 2600MC, 9-minute exposure. Halifax NS, Dec 9th 2023.
135: (Left) Askar 103 and ASI2600 camera, 6-minute exposure. Halifax, NS, May 13, 2025.
136: (Right) 20-second exposure, 2004.
137: Moon through Iphone. Halifax, Jan 25, 2018.

Made in United States
North Haven, CT
10 November 2026

22088883R00085